Lecture Notes in Mathematics

Volume 2321

This series reports on new developments in all areas of mathematics and their applications - quickly, informally and at a high level. Mathematical texts analysing new developments in modelling and numerical simulation are welcome. The type of material considered for publication includes:

1. Research monographs
2. Lectures on a new field or presentations of a new angle in a classical field
3. Summer schools and intensive courses on topics of current research.

Texts which are out of print but still in demand may also be considered if they fall within these categories. The timeliness of a manuscript is sometimes more important than its form, which may be preliminary or tentative.

Titles from this series are indexed by Scopus, Web of Science, Mathematical Reviews, and zbMATH.

Benno van den Berg • Eric Faber

Effective Kan Fibrations in Simplicial Sets

Benno van den Berg
Institute for Logic, Language and Computation
Universiteit van Amsterdam
Amsterdam, The Netherlands

Eric Faber
Frankfurt am Main, Germany

ISSN 0075-8434 ISSN 1617-9692 (electronic)
Lecture Notes in Mathematics
ISBN 978-3-031-18899-2 ISBN 978-3-031-18900-5 (eBook)
https://doi.org/10.1007/978-3-031-18900-5

Mathematics Subject Classification: 18N45, 18N50, 55U10, 55U35, 18C50

This Springer imprint is published by the registered company Springer Nature Switzerland AG
The registered company address is: Gewerbestrasse 11, 6330 Cham, Switzerland

Preface

This book starts to redevelop the foundations of simplicial homotopy theory, in particular around the Kan-Quillen model structure on simplicial sets, in a more effective or "structured" style. Our motivation comes from *homotopy type theory* (HoTT) and Voevodsky's construction of a model of HoTT in simplicial sets [1], which relies heavily on the existence and properties of the Kan-Quillen model structure.

Type theory refers to a family of formal systems which can act both as foundations for (constructive) mathematics and functional programming languages. Recently, it has become apparent that there exist many connections between type theory on the one hand and homotopy theory and higher category theory on the other. Besides Voevodsky's fundamental contributions, other key steps have been the groupoid model by Hoffman and Streicher [2], the interpretation of Martin-Löf's identity types in categories equipped with a weak factorisation system [3] and the proof that types in type theory carry the structure of an ∞-groupoid [4, 5]. As a result of these contributions, homotopy type theory has become an active area of research which keeps on developing at a quick pace, with implications for both type theory and homotopy theory. (For an overview, see the HoTT book [6].)

However, for type theory to fully benefit from the rich treasure chest of homotopy theory and higher category theory, a computational understanding of the relevant results from these areas is crucial. Indeed, we would like to think of type theory as a framework for computation. Then to fully exploit homotopy-theoretic ideas in this framework, one must be able to computationally reduce them. So a natural question is how constructive Voevodsky's model in simplicial sets is, or the proofs of the properties of the Kan-Quillen model structure on which it relies.

One fundamental obstacle with building Voevodsky's model in simplicial sets in a constructive framework was identified by Bezem, Coquand and Parmann [7]. To interpret Π-types in simplicial sets, one uses that the category of simplicial sets is locally Cartesian closed: that is, that the pullback functor along any map has a right adjoint, which we call *pushforward*. Since the type families are interpreted as Kan fibrations in Voevodsky's model, we need to show that Kan fibrations are closed under pushforward along Kan fibrations. This is true classically, but as the authors of [7] show, this result is unprovable constructively. We will refer to this as the BCP-obstruction and, given the importance of Π-types in type theory, it is quite a serious problem.

That this problem is not insurmountable was shown by Gambino and Sattler [8]: the key idea here is to treat being a Kan fibration not as a property, but as structure. Inspired by the work in HoTT on cubical sets ([9] in particular), they define a structured notion of a *uniform Kan fibration* and give a constructive proof that uniform Kan fibrations are closed under pushforward. They also show that their definition is "classically correct" in that a map can be equipped with the structure of a uniform Kan fibration if and only if it has the right lifting property against the horn inclusions (is a Kan fibration in the usual sense).

In this book, we will introduce another solution to this problem: the *effective Kan fibrations*. The reason for introducing a new solution is that Gambino and Sattler ran into trouble with another type constructor: universes. Indeed, the only known method for constructing universal fibrations in presheaf categories is via the Hofmann-Streicher construction [10], and this method can only be applied to notions of fibred structure which are *local* (see Definition 2.4 below). The problem is that the uniform Kan fibrations are not local, whereas we are able to show this for our notion of an effective Kan fibration. (That uniform Kan fibrations are not local was shown by Christian Sattler; his proof can be found in Appendix D in this book.)

So, to summarise, our main contribution is the introduction of the notion of an effective Kan fibration, a structured notion of Kan fibration for which we will prove the following results:

1. Effective Kan fibrations are closed under pushforward.
2. The notion of an effective Kan fibration is local and hence universal effective Kan fibrations exist.
3. Effective Kan fibrations have the right lifting property against horn inclusions.
4. A map which has the right lifting property against horn inclusions can be equipped with the structure of an effective Kan fibration.

We will give constructive proofs of (1)–(3), whereas the proof of (4) will necessarily be ineffective (due to the BCP obstruction). As a result, the effective Kan fibrations are the first, and so far only, structured notion of fibration for which these properties have been shown.

Besides having a clear computational content, another advantage of constructive proofs is that they can be internalised to arbitrary Grothendieck toposes (not just *Sets*). In fact, our arguments can be formalised in any locally Cartesian closed category with finite colimits and a natural numbers object. For those who prefer to think in terms of set theory, our arguments can be performed in (a subsystem of) Aczel's constructive set theory **CZF** (for which see [11]), which in turn is a subsystem of classical **ZF**, Zermelo-Fraenkel set theory (without choice).

But however this may be, we feel that laying too great an emphasis on the metamathematical aspects of our work may be misleading. The task of reworking some of the fundamental concepts in simplicial homotopy theory in a more explicit or structured style is an interesting undertaking in itself, whatever one's foundational convictions, and we hope that any homotopy theorists reading this work will come to see it that way as well. Indeed, any mathematician who wishes to skip the occasional foundational aside on our part should feel free to do so and can read this book as just another piece of new mathematics.

Related Work

Besides the work of Gambino and Sattler we already mentioned, there are two strands of research with which our approach should be compared.

In response to the BCP obstruction, most researchers in HoTT have abandoned simplicial sets and switched to cubical sets. In doing so, people have managed to constructively prove the existence of a model structure and a model of HoTT in cubical sets. In addition, their cubical models can be seen as interpreting a cubical type theory, in which principles like univalence can be derived and which enjoys (homotopy) canonicity (see [9, 12–14]).

These are impressive results and our approach is by no means that far advanced. However, we still feel that analogous results for simplicial sets would be preferable: indeed, simplicial techniques pervade modern homotopy theory, much more than cubical approaches do, and in order to connect to most of the ongoing work in homotopy theory and higher category theory, a simplicial approach is more likely to be successful. In addition, it is at present not entirely clear whether any of the constructive model structures that people have developed in cubical sets model the world of homotopy types or ∞-groupoids.

The other approach with which our work should be compared is that of Gambino, Henry, Sattler and Szumiło, who, in face of the BCP obstruction, decide to bite the bullet (see [15–17]). Their starting point was the constructive proof by Simon Henry of the existence of the Kan-Quillen model structure on simplicial sets, using the standard definitions of the Kan and trivial Kan fibrations (having the right lifting property against the horn inclusions and boundary inclusions, respectively). Based on this work, Henry in collaboration with Gambino managed to construct a model of HoTT, modulo some tricky coherence problems. Their work has the advantage that it is based on the usual definitions of the (trivial) Kan fibrations, so in that sense it looks, at least at first glance, more familiar than our structured approach. In addition, their work is definitely more advanced than ours.

However, we still think that a structured approach looks more appealing. Due to the BCP obstruction, they only have a weak form of Π-types. In comparision, our approach should give us genuine Π-types with definitional η- and β-rules. Also, it seems that to obtain a genuine model of homotopy type theory based on their work forces one to solve some quite difficult coherence problems, for which at present no solutions are known. In contrast, we expect that a more structured approach will be helpful in solving any coherence problems we would encounter if we would start to turn our work into a model of type theory.

Acknowledgements

We thank Richard Garner for some very useful conversations on polynomial functors, which had a major influence on Chap. 9. We also thank Christian Sattler for allowing us to include his proof about the non-locality of the uniform Kan fibrations. Finally, we are also grateful to the referees for very helpful feedback.

Amsterdam, The Netherlands Benno van den Berg
Frankfurt am Main, Germany Eric Faber

Contents

Chapter 1
Introduction

1.1 Fibrations as Structure

The main contribution of this book is the introduction of the concept of an effective Kan fibration. So what is our notion of an effective Kan fibration? Before we answer that question, let us first discuss what, in general, we mean by a structured notion of fibration.

A common situation in homotopy theory is that we are working in some category $\mathcal{E}$ equipped with a pullback stable class of fibrations; by pullback stability we mean that in a pullback square like

$$\begin{array}{ccc} Y' & \longrightarrow & Y \\ \downarrow{\scriptstyle p'} & & \downarrow{\scriptstyle p} \\ X' & \longrightarrow & X \end{array}$$

the map p' will be a fibration whenever p is. If one conceptualises matters like this, being a fibration is a *property* of a map; in this book, however, we will think of being a fibration as *structure*. In particular, the setting of a category $\mathcal{E}$ with a pullback stable class of fibrations will be replaced by a *structured notion of a fibration* or *a notion of fibred structure* on $\mathcal{E}$, which we will define as a presheaf

$$\mathrm{Fib} : (\mathcal{E}^{\to}_{\mathrm{cart}})^{op} \to \mathit{Sets},$$

where $\mathcal{E}^{\to}_{\mathrm{cart}}$ denotes the category of arrows in $\mathcal{E}$ and pullback squares between them. Given such a structured notion of fibration, an element $\sigma \in \mathrm{Fib}(p)$ will be called a *fibration structure* on $p : Y \to X$; and if such an element σ exists, we may call p a *fibration*. These fibrations will form a pullback stable class, as before. Indeed, we can think of a pullback stable class as a degenerate notion of fibred structure

B. van den Berg, E. Faber, *Effective Kan Fibrations in Simplicial Sets*,
Lecture Notes in Mathematics 2321, https://doi.org/10.1007/978-3-031-18900-5_1

where Fib(p) always contains at most one element, signalling whether the map p is a fibration or not.

But in many examples being a fibration is quite naturally thought of as additional structure on a map. For instance, a common way of defining a class of fibrations is by saying that they are *cofibrantly generated* by a class of maps $\mathcal{A}$. That is, a map $p : Y \to X$ is a fibration precisely when for any $m : B \to A \in \mathcal{A}$ and commutative square

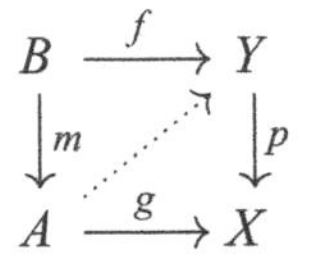

$$\begin{array}{ccc} B & \xrightarrow{f} & Y \\ {\scriptstyle m}\downarrow & \nearrow & \downarrow{\scriptstyle p} \\ A & \xrightarrow{g} & X \end{array}$$

there exists a dotted filler as shown making both triangles commute; one also says that p has the *right lifting property* against $\mathcal{A}$. In this situation we can define a structured notion of fibration Fib by declaring the elements of Fib(p) to be *lifting structures* on p: that is, functions which assign to each square like the one above with $m \in \mathcal{A}$ a filler $\sigma_{m,f,g} : A \to Y$ making both triangles commute. Let us for the moment write this notion of fibred structure as RLP$(\mathcal{A})$.

One thing which happens if one shifts to a structured style is that notions of fibration which are *logically equivalent* as properties are no longer *isomorphic* as structures. Take the trivial Kan fibrations in simplicial sets as an example. They can be defined as the maps which are cofibrantly generated by the monomorphisms; or as those which are cofibrantly generated by the monomorphisms $S \subseteq \Delta^n$ with representable codomain; or as those which are cofibrantly generated by the boundary inclusions $\partial\Delta^n \subseteq \Delta^n$. These may all be equivalent as properties, but as structures, they are all different. Indeed, there are "forgetful" morphisms of fibred structure (presheaves)

$$\text{RLP(monos)} \to \text{RLP(sieves)} \to \text{RLP(boundary inclusions)},$$

but they are not monomorphisms, let alone isomorphisms. So as structured notions of fibration they need to be carefully distinguished.

This leads us to another important point for this book: one can try to repair this by imposing *compatibility conditions* on the lifting structure (also known as *uniformity conditions* in the literature). Indeed, in the way we have defined RLP$(\mathcal{A})$ its elements σ choose solutions for a class of lifting problems, but there are no conditions saying how these solutions should be related. For instance, suppose we have in simplicial sets a solid diagram of the form

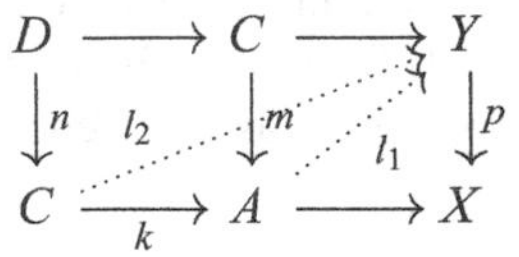

$$\begin{array}{ccccc} D & \longrightarrow & C & \longrightarrow & Y \\ {\scriptstyle n}\downarrow & {\scriptstyle l_2} & \downarrow{\scriptstyle m} & {\scriptstyle l_1} & \downarrow{\scriptstyle p} \\ C & \xrightarrow[k]{} & A & \longrightarrow & X \end{array}$$

in which p is a trivial Kan fibration and the left hand square is a pullback involving monomorphisms n and m. Then any element in RLP(monos)(p) must, among other things, choose dotted arrows l_1 and l_2 as shown; we could define a notion of fibred structure RLP_c(monos) which would require that in such circumstances we must have $k.l_1 = l_2$. And if RLP_c(sieves) would be the restriction of RLP(sieves) to those lifting structures which in a situation like

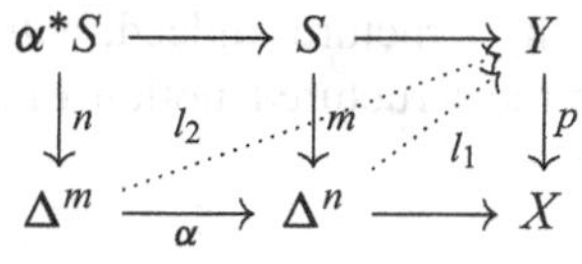

$$\begin{array}{ccccc} \alpha^*S & \longrightarrow & S & \longrightarrow & Y \\ \downarrow n & l_2 & \downarrow m & l_1 & \downarrow p \\ \Delta^m & \xrightarrow[\alpha]{} & \Delta^n & \longrightarrow & X \end{array}$$

would choose lifts l_1 and l_2 satisfying $\alpha.l_1 = l_2$, then the forgetful morphism

$$\mathrm{RLP}_c(\text{monos}) \to \mathrm{RLP}_c(\text{sieves})$$

would be an isomorphism of notions of fibred structure. The reader may wonder if any trivial Kan fibration can still be equipped with such a structure: that is, whether lifts against monos or sieves can always be chosen in such a way that these compatibility conditions are met. That is indeed the case (see [8]).

Both RLP_c(monos) and RLP_c(sieves) are examples of right lifting structures defined by lifts against *categories* rather than *classes* of maps, and a rich variety of lifting structures is quite characteristic of our structured approach. Indeed, we will also consider *double categorical* and even *triple categorical* notions of lifting structure. To motivate this, let us consider the forgetful map

$$\mathrm{RLP}(\text{monos}) \to \mathrm{RLP}(\text{boundary inclusions}).$$

This will not be a monomorphism even when we restrict to RLP_c(monos), but there is a further (double-categorical) compatibility condition we could imagine imposing which would have this effect. Suppose we have a solid diagram in simplicial sets

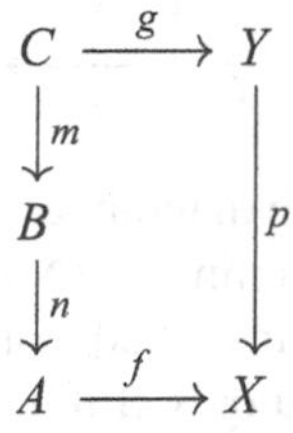

$$\begin{array}{ccc} C & \xrightarrow{g} & Y \\ \downarrow m & & \\ B & & \downarrow p \\ \downarrow n & & \\ A & \xrightarrow{f} & X \end{array}$$

in which p is a trivial Kan fibration and m and n are monomorphisms. Then a lifting structure σ on p will give rise to a filler $A \to Y$ in two different ways: we can use that monomorphisms are closed under composition and take $\sigma_{n.m,g,f}$. But we could also first construct a map $l : B \to Y$ by taking $l = \sigma_{m,g,f.n}$ and then use that to construct $\sigma_{n,l,g}$. A natural requirement would be that these two lifts should always coincide. If we write RLP_{dc}(monos) for the notion of fibred structure where the lifts satisfy this condition on top of the previous one, then we will prove in this book that

$$\mathrm{RLP}_{dc}(\text{monos}) \to \mathrm{RLP}(\text{boundary inclusions}).$$

is a monomorphism of notions of fibred structure. We will also characterise the image of this map and show that every trivial Kan fibration can be equipped with such a double-categorical lifting structure. Indeed, with one further (constructive) twist, this will be our preferred structured notion of a trivial Kan fibration (an *effective trivial Kan fibration*).

1.2 Effective Kan Fibrations

As said, the core of our book is the definition of an effective Kan fibration, our preferred structured notion of a Kan fibration. To motivate this definition, let us recall the classical result (from [18]) that says that the Kan fibrations are cofibrantly generated by maps of the form $m \hat{\otimes} \partial_i$, where $m : A \to B$ is a cofibration, $\hat{\otimes}$ is the pushout-product and $\partial_i : 1 \to \mathbb{I}$ is one of the two endpoint inclusions into the interval $\mathbb{I} = \Delta^1$. This can be reformulated as follows: let us say that a map $p : Y \to X$ has the right lifting property against a commutative square

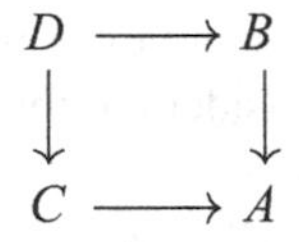

if for any solid diagram

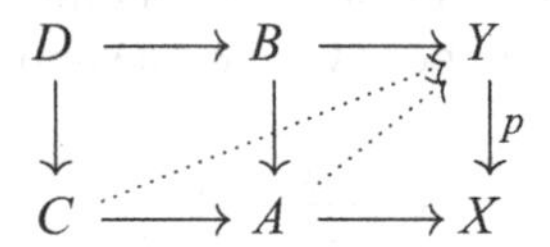

and dotted arrow $C \to Y$ making the diagram commute, there exists a dotted arrow $A \to Y$ making the whole picture commute. (Note that this is equivalent to having the right lifting property against the inscribed map from the pushout $B \coprod_D C$ to A.) Then a map is a Kan fibration if and only if it has the right lifting property against the left hand square in a double pullback diagram of the form

$$\begin{array}{ccccc} A & \xrightarrow{(1,\partial_i)} & A \times \mathbb{I} & \xrightarrow{\pi_1} & A \\ \downarrow{\scriptstyle m} & & \downarrow{\scriptstyle m\times\mathbb{I}} & & \downarrow{\scriptstyle m} \\ B & \xrightarrow{(1,\partial_i)} & B \times \mathbb{I} & \xrightarrow{\pi_1} & B. \end{array}$$

The usual definition of a Kan fibration in terms of horn inclusions can also be stated as a lifting problem against a square, namely the left hand square in another double pullback diagram:

$$\begin{array}{ccccc} \partial\Delta^n & \longrightarrow & s^*\partial\Delta^n & \longrightarrow & \partial\Delta^n \\ \downarrow & & \downarrow & & \downarrow \\ \Delta^n & \xrightarrow{d} & \Delta^{n+1} & \xrightarrow{s} & \Delta^n, \end{array}$$

where $s = s_i$ is one of the degeneracies and $d = d_i/d_{i+1}$ is one of its sections. (The reason being that the inclusion $\Delta^n \cup s^*\partial\Delta^n \to \Delta^{n+1}$ is the horn inclusion $\Lambda^{n+1}_{i/i+1}$.) We will call such left hand squares *horn squares*.

These two situations have something in common, namely that they are both lifting conditions against a left hand square in a double pullback diagram of the form

$$\begin{array}{ccccc} C & \longrightarrow & r^*C & \longrightarrow & C \\ \downarrow{\scriptstyle m} & & \downarrow & & \downarrow{\scriptstyle m} \\ A & \xrightarrow{i} & B & \xrightarrow{r} & A \end{array}$$

in which m is a cofibration and (i, r) is a deformation retract of some kind. The first step towards our definition of an effective Kan fibration is the identification of the right kind of deformation retracts. Our solution is the notion of a *hyperdeformation retract* (HDR), and to define these HDRs we use the simplicial Moore path functor defined by the first author in collaboration with Richard Garner [19].

Once we have the concept of an HDR, we can define the *mould squares* as those pullback squares

$$\begin{array}{ccc} A' & \xrightarrow{(i',r')} & B' \\ \downarrow{\scriptstyle m} & & \downarrow \\ A & \xrightarrow{(i,r)} & B \end{array}$$

in which m is a cofibration, (i, r) and (i', r') are HDRs and the square (read from top to bottom) is what we will call a cartesian morphism of HDRs. The idea, then, is to define the effective Kan fibrations as those maps which come equipped with lifts against mould squares.

What is missing from this definition, however, are the correct compatibility conditions. It turns out that mould squares can be composed both horizontally and vertically (they naturally fit into a double category), and this leads to two natural compatibility conditions. In fact, there is a further "perpendicular" condition, because mould squares can be pulled back along morphisms of HDRs, leading to a third (triple-categorical) compatibility condition. Indeed, our main reason for introducing the concept of a mould square is that they allow us to express these compatibility conditions in such an elegant way.

Box 1.1 From Uniform to Effective Kan Fibrations

One way of understanding our notion of an effective Kan fibration is as a modification of the notion of a uniform Kan fibration by Gambino and Sattler. They work with a category of cofibrations and an interval object $\mathbb{I}$: in the category of simplicial sets these would be a category of monomorpisms and pullback squares between them and the representable $\mathbb{I} = \Delta^1$. The interval comes equipped with two maps $\partial_0, \partial_1 : 1 \to \mathbb{I}$ which yields two natural transformations $s_Y, t_Y : Y^{\mathbb{I}} \to Y$. A map $p : Y \to X$ is a uniform Kan fibration if $(s/t, p^{\mathbb{I}}) : Y^{\mathbb{I}} \to Y \times_X X^{\mathbb{I}}$ has the right lifting property against the category of cofibrations. We modify this definition by replacing the path object $X^{\mathbb{I}}$ by the simplical Moore path object MX and adding a uniformity condition demanding that lifts behave well with respect to concatenation of Moore paths, which implies that lifts of general simplicial Moore paths are determined by those of length 1. This modification can be understood as a consequence of our desire to make the definition local. Indeed, note that an n-simplex in $Y^{\mathbb{I}}$ corresponds to a prism $\Delta^n \times \mathbb{I} \to Y$. In contrast with cubical sets, the representable simplicial sets are not closed under products and therefore the domain of this map is not representable. This is ultimately what causes non-locality of the uniform Kan fibrations. However, the map $\Delta^n \times \mathbb{I} \to Y$ can be understood as a Moore path of length $n+1$, composed of Moore paths of length one of the form $\Delta^{n+1} \to Y$, reflecting the fact that the prism $\Delta^n \times \mathbb{I}$ is the union of $(n+1)$-many $(n+1)$-simplices. Therefore for effective Kan fibrations all the structure is determined by the lifts for maps with representable domain and this is what allows us to regain locality.

We will also add a further compatibility condition for our effective Kan fibrations which relates to the fact that cofibrations are closed under composition. This additional requirement is not strictly necessary for our purposes, but it allows us to prove in Chap. 12 that the structure of being an effective Kan fibration is completely determined by the lifts against the horn inclusions (or horn squares, to be more precise). In addition, as we will show in future work, it will have the consequence that the effective trivial fibrations and the effective Kan fibrations interact nicely.

Once we have this in place, and we have checked that horn squares are mould squares, it follows immediately that effective Kan fibrations have the right lifting property against horn inclusions (it will also not be too hard to see that our effective Kan fibrations are uniform Kan fibrations in the sense of Gambino-Sattler). In fact, quite a lot of pages will be spent on proving that the lifts against the mould squares are completely determined by the lifts against the horn squares, or, in other words, that the forgetful map

$$\text{RLP}_{tc}(\text{mould squares}) \to \text{RLP}(\text{horn squares})$$

is a monomorphism of fibred structures. We will also characterise its image, which will be crucial for proving both that our notion of an effective Kan fibration is local and that it is classically correct. This book consist of two parts, and these two results will form the main achievements of the second part.

The first part will be devoted to proving that the effective Kan fibrations are closed under pushforward. We find it convenient to do this axiomatically, using an axiomatic setup reminiscent of the work of Orton and Pitts [20]. The idea of Orton and Pitts (but see also [8, 21]) was to develop the basic theory of the cubical sets model in the setting of a suitable category equipped with a class of cofibrations forming a dominance and an interval object $\mathbb{I}$. In our setup we will keep the dominance, but replace the interval object by a Moore path functor M satisfying certain equations (these can be found in Appendix A in this book). The example we have in mind is, of course, the simplicial Moore path functor from [19]. As our dominance, we take the monomorphisms in simplicial sets which are "pointwise decidable" (this is an additional constructive requirement that we impose on the cofibrations, which can be ignored by our classical readers). As we will show, this axiomatic setting is sufficiently powerful to define a suitable notion of mould square and effective Kan fibration, and prove that the effective Kan fibrations are closed under pushforward.

The main drawback of the notion of an effective Kan fibration might be that we are unable to constructively prove that they are closed under retracts. That is, we do not know how to effectively equip maps which are retracts of effective Kan fibration with the structure of an effective Kan fibration. In fact, we suspect that this is impossible, but we do not know this for sure.

The other drawback is that the theory is currently not complete. It should be the case that the effective Kan fibrations are the right class in an algebraic weak factorisation system and that they underlie an algebraic model structure as in [22]. We plan to take this up in future work.

1.3 Summary of Contents

The contents of this book are therefore as follows.

We start Part I with a recap of the theory of algebraic weak factorisation systems (AWFSs), a structured analogue of the notion of a weak factorisation system. In this structured notion the left maps are replaced by coalgebras for a comonad on the arrow category, while the right maps are replaced by the algebras for a monad on the arrow category. Our main reference for this theory is an important paper by Bourke and Garner [23], which also explains the connection to double categories. There are two (related) points here which are perhaps worth stressing for those who are already familiar with this theory: first of all, for us the distributive law is important and we will always assume it. Secondly, we will exclusively work with the (co)algebras for the (co)monad, never with the (co)algebras for the (co)pointed endofunctor. The reason is that we will not assume that the effective Kan fibrations are closed under retracts and therefore we cannot express them as algebras for a pointed endofunctor. It also means that we cannot expect them to be cofibrantly generated by a small category: the best we can hope for is that they are cofibrantly generated by a small double category. (We believe this to be true, but we will not prove this in this book.) In any case, the connection of algebraic weak factorisation systems to double categories is quite important for us. But to make that work, the distributive law is crucial and that is the reason we will always assume it.

We will then go on to explain how both dominances and Moore structures give rise to AWFSs (for dominances this can already be found in [23]). We will refer to these as the (effective cofibration, effective trivial fibration) and (HDR, naive fibration)-AWFS, respectively. Using these two ingredients we will then define the notions of mould square and effective fibration. Assuming that the Moore structure is symmetric, we will then show that these effective fibrations are closed under pushforward. An important intermediate step for this is the proof of the Frobenius property for the (HDR, naive fibration)-AWFS, which is related to an argument that can also be found in [19].

We will start Part II by showing that the category of simplicial sets can be equipped with both a dominance and a symmetric Moore structure. This will show that the theory of Part I applies to simplicial sets. Then we will proceed to show that effective (Kan) fibrations can be completely characterised by their lifts against horn squares, which will prove both that this notion of fibred structure is local and classically correct.

To our surprise it turns out that the machinery we develop here can also be used to give effective (structured) analogues of the notions of left and right fibration in simplicial sets. Indeed, also these can be defined by a right lifting property against a class of mould squares, using the same dominance of effective cofibrations, but a different Moore structure. When our results have implications for an effective theory of left and right fibrations, we will comment on that as well.

Finally, we will finish this book with a conclusion outlining directions for future research and four appendices. In the first appendix we give our version of the Orton-

Pitts axioms. In the second appendix we compare the axioms for Moore structure with the counterpart notions of connections and diagonals in cubical sets. The third appendix proves a result on horn fillers that we need for the proofs that our different effective notions of fibration are classically correct. Finally, the fourth appendix contains Christian Sattler's proof that the notion of a *uniform* Kan fibration in the category of simplicial sets is not local.

Part I
Π-Types from Moore Paths

Chapter 2
Preliminaries

In this chapter we introduce the main theoretical framework in which our theory of effective fibrations is embedded. Abstractly put, we are studying and constructing new notions of *fibred structure* and *cofibred structure* on a category $\mathcal{E}$.

Throughout the first part, we will assume that $\mathcal{E}$ is a locally cartesian closed category with finite limits and colimits. Certain results may also be obtained for a wider class of categories $\mathcal{E}$. For example, those results which only use that the codomain functor is a bifibration satisfying the Beck-Chevalley condition, hold more generally than just for locally cartesian categories with finite colimits, because the latter notion is stronger (Lemma 2.1). Yet in other chapters we do resort to the fact that $\mathcal{E}$ is locally cartesian closed, and hence stating results which combines the theory from different chapters quickly becomes unwieldy when working in full generality. In short, the theory presented in this book is best understood when one is not so much worried about its level of generality.

We first recall the definition of a locally cartesian closed category.

Definition 2.1 Suppose $\mathcal{E}$ is a category with pullbacks. Then $\mathcal{E}$ is *locally cartesian closed* if for every arrow $f : X \to Y$, the pullback functor

$$f^* : \mathcal{E}/Y \to \mathcal{E}/X \tag{2.1}$$

has a right adjoint:

$$\Pi_f : \mathcal{E}/X \to \mathcal{E}/Y. \tag{2.2}$$

Instead of Π_f we will also often write f_* and we refer to this functor as the *pushforward along* f.

Definition 2.2 (Bénabou et al. [24]) A functor $F : \mathcal{F} \to \mathcal{E}$ is *bifibration* when it is a Grothendieck fibration as well as an opfibration.

B. van den Berg, E. Faber, *Effective Kan Fibrations in Simplicial Sets*,
Lecture Notes in Mathematics 2321, https://doi.org/10.1007/978-3-031-18900-5_2

In the first part, most notably in Chap. 4, we use the following important property of a locally cartesian closed category with finite colimits. The functor

$$\mathrm{cod} : \mathcal{E}^{\rightarrow} \rightarrow \mathcal{E} \tag{2.3}$$

from the arrow category of $\mathcal{E}$ to $\mathcal{E}$ which sends arrows to their codomain is a bifibration satisfying the *Beck-Chevalley condition* (see Box 2.1). Besides pullbacks (which we already have), this also requires the existence of arbitrary pushouts, together with a compatibility condition between them. For (2.3), this condition is as follows. Given a commutative cube:

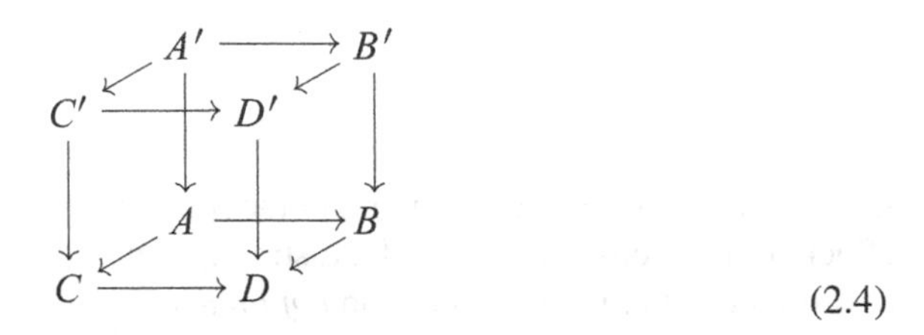

(2.4)

such that

(i) The bottom square $ABCD$ is a pullback;
(ii) The right square $B'D'BD$ is a pushout;
(iii) The back square $A'B'AB$ is a pullback;

then the left square $A'C'AC$ is a pushout if and only if the front square $C'D'CD$ is a pullback. In many cases, we obtain this condition from the following lemma:

Lemma 2.1 *If $\mathcal{E}$ is locally cartesian closed with finite colimits, then the codomain bifibration satisfies the Beck-Chevalley condition. That is, for all cubes* (2.4) *satisfying (i)–(iii), the above compatibility condition holds.*

Proof With pullbacks and pushouts present, it is easy to see that cod is in fact a bifibration. For bifibrations, it is enough to only show one direction of the compatibility assertion, for which see Box 2.1 on p. 15. Hence it suffices to show that when the front square is a pullback square, the left square is a pushout.

In a locally cartesian closed category, pullback along any $f : C \rightarrow D$ has a right adjoint and therefore preserves colimits. Since the right-hand square is a pushout, it is also a pushout in $\mathcal{E}/D$ and hence preserved by pullback along f. Hence, as the front square is in fact a pullback, the left square is a pushout in $\mathcal{E}/C$. It follows that it is also a pushout square in $\mathcal{E}$. This proves the assertion. □

Box 2.1 The Beck-Chevalley Condition for Bifibrations
The following definition of the *Beck-Chevalley condition* for bifibrations originates from Bénabou-Roubaud [24]. For a bifibration $\mathcal{F} \to \mathcal{E}$, the condition is satisfied when, as drawn in the diagram below, for every commutative square in the fibre above a pullback square

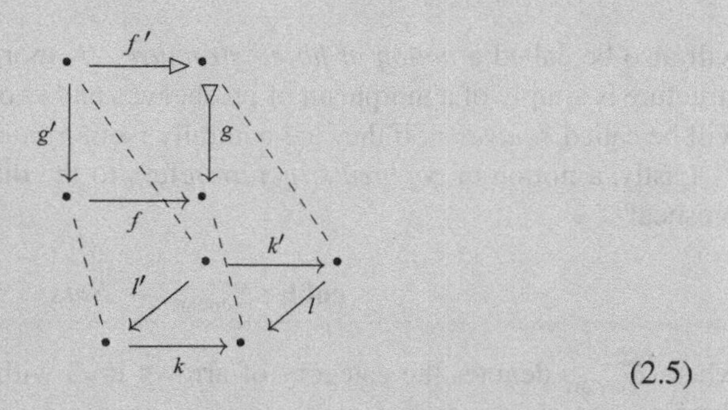

(2.5)

such that f' is cartesian and g is cocartesian, one has that f is cartesian if and only if g' is cocartesian. Note that in fact it is equivalent that only one of these two directions hold: this can be seen by factorizing the diagonal either as a precomposition with a cocartesian arrow, or a postcomposition with a cartesian arrow.

If the bifibration comes with a choice of cartesian and cocartesian lifts in the form of fibrewise pullback $(-)^*$ and pushforward $(-)_*$, this can be written as an isomorphism:

$$l'_* k'^* \cong k^* l_*$$

for every pullback square in the base as drawn.

Another property that we need later (specifically in the proof of Proposition 6.1) is that any locally cartesian closed category with finite colimits is *coherent*. Recall that a *coherent category* is a category where subobjects have finite unions preserved by pullback. For this to hold, it is enough that this category has finite coproducts and is *regular* (see Johnstone [25], A1.4).

Lemma 2.2 *A locally cartesian closed category $\mathcal{E}$ with finite colimits is coherent.*

Proof By the preceding paragraph, it is enough to show that $\mathcal{E}$ is a regular category. This follows from Johnstone [25], A1.3.5, where we use that pullback is a left adjoint and hence preserves colimits. □

2.1 Fibred Structure

Definition 2.3 Let $\mathcal{E}$ be a category with finite limits and write $\mathcal{E}^{\to}_{\text{cart}}$ for the category of arrows in $\mathcal{E}$ and pullback squares between them. A presheaf **fib** on $\mathcal{E}^{\to}_{\text{cart}}$

$$\mathbf{fib} : (\mathcal{E}^{\to}_{\text{cart}})^{\text{op}} \to \mathcal{S}ets$$

will also be called a *notion of fibred structure* . A morphism of notions of fibred structure is simply of a morphism of presheaves and two notions of fibred structure will be called *equivalent* if they are naturally isomorphic as presheaves.

Lastly, a notion of *cofibred structure* refers to the dual, which is the same as a presheaf

$$\mathbf{cofib} : \mathcal{E}^{\to}_{\text{cocart}} \to \mathcal{S}ets$$

where $\mathcal{E}^{\to}_{\text{cocart}}$ denotes the category of arrows in $\mathcal{E}$ with pushout squares between them.

Definition 2.4 Let **fib** be a notion of fibred structure on a category $\mathcal{E}$. We will call the notion of fibred structure **fib** *local* (or *locally representable*) if the following holds for any small diagram $D : \mathcal{I} \to \mathcal{E}^{\to}_{\text{cart}}$ with colimit f and colimiting cocone $(\sigma_i : Di \to f : i \in \text{Ob}(\mathcal{I}))$:

> If we can choose fibration structures $x_i \in \mathbf{fib}(Di)$ for any $i \in \text{Ob}(\mathcal{I})$ such that $\mathbf{fib}(D\alpha)(x_i) = x_j$ for any $\alpha : j \to i$ in $\mathcal{I}$, then there exists a unique fibration structure $x \in \mathbf{fib}(f)$ such that $\mathbf{fib}(D\sigma_i)(x) = x_i$.

We will mainly be interested in notions of fibred structure on presheaf categories (in fact, on the category of simplicial sets), in which case the notion of locality can be defined in a different way. In the following proposition, the Yoneda embedding for a category of presheaves $\mathcal{E}$ on $\mathbb{C}$ is denoted $y_{(-)} : \mathbb{C} \to \mathcal{E}$.

Proposition 2.1 *Suppose $\mathcal{E}$ is the category of presheaves on $\mathbb{C}$, and let* **fib** *be a notion of fibred structure on $\mathcal{E}$. Then* **fib** *is local if and only if the following holds for any morphism $f : Y \to X$ in $\mathcal{E}$: given, for any $x \in X(C)$, a chosen fibration structure s_x on a (chosen) pullback $f_x : Y_x \to y_C$ as in*

$$\begin{array}{ccc} Y_x & \longrightarrow & Y \\ \big\downarrow{\scriptstyle f_x} & & \big\downarrow{\scriptstyle f} \\ y_C & \xrightarrow{x} & X \end{array}$$

such that for any $x \in X(C)$ and $\alpha : D \to C$ in $\mathbb{C}$, the fibration structure s_x on f_x pulls back to the one chosen on $f_{x\cdot\alpha}$ for the pullback square

$$\begin{array}{ccc} Y_{x\cdot\alpha} & \longrightarrow & Y_x \\ \big\downarrow{\scriptstyle f_{x\cdot\alpha}} & & \big\downarrow{\scriptstyle f_x} \\ yD & \xrightarrow{\alpha} & yC, \end{array}$$

over f, then there exists a unique fibration structure $s \in \mathbf{fib}(f)$ which pulls back to s_x for any pullback square of the first type.

Proof The equivalence uses standard properties of presheaf categories: every object is a colimit of representables, and since pullback along f preserves colimits, the condition becomes a special case of Definition 2.4.

For the other direction, one can reduce Definition 2.4 to the special case by taking pullbacks along representables:

$$\begin{array}{ccccc} Y_{i,x} & \longrightarrow & Y_i & \longrightarrow & Y \\ \big\downarrow & & \big\downarrow{\scriptstyle D(i)} & \overset{\sigma_i}{\Rightarrow} & \big\downarrow{\scriptstyle f} \\ yC & \xrightarrow[x]{} & X_i & \longrightarrow & X \end{array}$$

Since $\mathrm{cod} : \mathcal{E}^{\to} \to \mathcal{E}$ is a left adjoint, it preserves colimits, so every $x' : y_C \to X$ factors through some X_i since we are in a category of presheaves. Hence the fibred structure determines precisely the input data for the special case. The pullback property for the unique induced fibration follows from the uniqueness condition on each of the $D(i)$ that follows from the special case. □

Remark 2.1 A local notion of fibred structure is a structured analogue of a local class of maps as in [26, Remark 4.4] and [27, Definition 3.7], for instance. An earlier structured analogue appears in Shulman's paper [28, Proposition 3.18]. The definition we gave here is a special case of his, because we demand that the fibration structures are strictly functorial under pullback, rather than pseudofunctorial.

A motivation behind a local notion fibred structure that is of interest to type theory is that it allows to give a *universal* fibration. For this to work we assume that we are given some notion of "smallness" in the metatheory (like being finite, or countable, or being bound in size by some regular cardinal, or belonging to some Grothendieck universe). If $\mathcal{E}$ is the category of presheaves on a *small* category $\mathbb{C}$, then we will call a morphism of presheaves $p : Y \to X$ *small* if each fibre of $p_C : Y(C) \to X(C)$ is small. A presheaf X will be called small if $X \to 1$ is a small morphism of presheaves.

Using the same notation as in Proposition 2.1, we have the following theorem:

Theorem 2.1 *Suppose $\mathcal{E}$ is the category of presheaves on a small category $\mathbb{C}$, and let* **fib** *be a local notion of fibred structure on $\mathcal{E}$.*

Then, there is a small morphism of presheaves $\pi : E \to U$ equipped with a unique fibration structure $s_\pi \in \mathbf{fib}(\pi)$ such that every small $p : Y \to X$ equipped with a structure of fibration $s_p \in \mathbf{fib}(p)$ can be uniquely presented as a pullback of π with this structure.

Proof The idea is to modify the Hofmann-Streicher construction as in [10]. That is, we define

$$U(C) = \{(p, s_p) \mid p : X \to y_C \text{ small, and } s_p \in \mathbf{fib}(p)\}$$
$$E(C) = \{(x, p, s_p) \mid p : X \to y_C \text{ small, } s_p \in \mathbf{fib}(p)\,, x \in X(C)\}$$

and for any $\alpha : D \to C$, maps

$$U(\alpha) : U(C) \to U(D)$$
$$E(\alpha) : E(C) \to E(D)$$

by pullback:

$$U(\alpha)(p, s_p) = ({y_\alpha}^* p, \mathbf{fib}({y_\alpha}^* p)(s_p))$$
$$E(\alpha)(x, p, s_p) = (X(\alpha)(x), {y_\alpha}^* p, \mathbf{fib}({y_\alpha}^* p)(s_p)).$$

We claim that the projection $\pi : E \to U$ has a fibration structure. Indeed, this follows from the local property of **fib**: for every pullback diagram

$$\begin{array}{ccc} X & \longrightarrow & E \\ {\scriptstyle p}\downarrow & & \downarrow{\scriptstyle \pi} \\ y_C & \xrightarrow{(p, s_p)} & U \end{array}$$

s_p is a natural choice of fibration structure on p that satisfies the condition in Proposition 2.1. By the Yoneda lemma, every arrow $u : y_C \to U$ is of this type. Hence there is a unique fibration structure on π such that for every $p : X \to y_C$, any fibration structure on p can be presented as a pullback of π in the way shown. This immediately generalises to any small morphism $p : Y \to X$, using again that **fib** is local.

The construction of E and U as we have just presented them does not quite work, because the action $U(\alpha)$ and $E(\alpha)$ is not strictly functorial. However, we can solve this issue by replacing the codomain fibration over $\mathcal{E}$ by an equivalent strict fibration. For instance, we can use the fact that the slice category of the category of presheaves over P is equivalent to the category of presheaves over $y \downarrow P$, the category of elements of P; this allows us to replace the codomain fibration over $\mathcal{E}$ by an equivalent split fibration whose fibre over a presheaf P in the base is the category of presheaves over $y \downarrow P$. □

2.2 Double Categories of Left and Right Lifting Structures

We recall the definition of a *double category*:

Definition 2.5 A *double category* $\mathbb{A}$ consists of:

(i) A collection of objects together with two separate category (morphism) structures on it, called *horizontal* and *vertical* morphisms.
(ii) A special category structure whose objects are the vertical morphisms, and whose arrows are called *squares*. The special property is that every *square* from a vertical morphism u to another vertical morphism v has a 'pointwise' domain, and a 'pointwise' codomain given by horizontal morphisms:

$$f : \operatorname{dom} u \to \operatorname{dom} v,\ g : \operatorname{cod} u \to \operatorname{cod} v$$

Moreover, composition of squares respects composition of these horizontal morphisms and identity squares have identity horizontal morphisms for their pointwise domain and codomain.

Further, there is a 'pointwise' or 'vertical' composition operation of squares with matching pointwise domains and codomains, which extends composition of vertical morphisms. We often think of a square $s : u \to v$ as filling in a diagram of horizontal and vertical arrows:

$$\begin{array}{ccc} A & \xrightarrow{f} & B \\ {\scriptstyle u}\downarrow & \overset{s}{\Rightarrow} & \downarrow{\scriptstyle v} \\ C & \xrightarrow[g]{} & D \end{array}$$

These diagrams can be composed both horizontally (ordinary composition of squares) and vertically (pointwise composition).

Alternatively, one can verify that a double category is the same thing as an internal category in the 'category' of large categories:

$$\mathbb{L}_1 \times_{\mathbb{L}_0} \mathbb{L}_1 \xrightarrow{\ \circ\ } \mathbb{L}_1 \begin{array}{c}\xrightarrow{\text{dom}} \\ \xleftarrow{\text{id}} \\ \xrightarrow[\text{cod}]{}\end{array} \mathbb{L}_0$$

where '$\circ$' denotes vertical, or pointwise, composition, whilst horizontal composition is the composition of the two categories $\mathbb{L}_1$ (vertical morphisms and squares) and $\mathbb{L}_0$ (objects and horizontal morphisms).

For double categories $\mathbb{A}$, $\mathbb{B}$, a *double functor* $F : \mathbb{A} \to \mathbb{B}$ is a compatible triple of functors (which we can denote by the same F) between categories of horizontal morphisms, vertical morphisms, and squares, which in addition respects pointwise composition. Here *compatible* means that the image of a square as drawn above looks like:

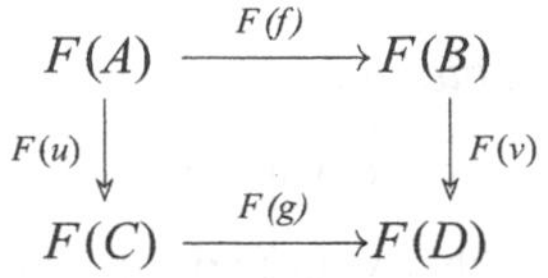

$$\begin{array}{ccc} F(A) & \xrightarrow{F(f)} & F(B) \\ {\scriptstyle F(u)}\downarrow & & \downarrow{\scriptstyle F(v)} \\ F(C) & \xrightarrow{F(g)} & F(D) \end{array}$$

Again, one can verify that a double functor between double categories is the same thing as an internal functor between the corresponding internal categories in the category of large categories.

Example: Double Category of Arrows

The typical example of a double category is the category of arrows of any category $\mathcal{E}$. The horizontal and vertical arrows are both given by the category structure, and squares are given by commutative squares. We will denote this double category by $\mathbf{Sq}\,(\mathcal{E})$.

In this book, we often work with double categories over the category of squares of some category $\mathcal{E}$, i.e. double functors $\mathbb{A} \to \mathbf{Sq}\,(\mathcal{E})$. For the next construction, we begin with such a double functor, denoted $I : \mathbb{L} \to \mathbf{Sq}\,(\mathcal{E})$. For a morphism $p : Y \to X$ in $\mathcal{E}$, a *right lifting structure* with respect to I consists of:

(i) A family $\phi_{-,-}(-)$ of arrows in $\mathcal{E}$ consisting of the following. For every vertical morphism v in $\mathbb{L}$, and every commutative square in $\mathcal{E}$, as in the solid part of the following diagram:

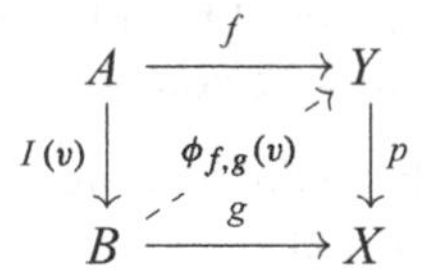

$$\begin{array}{ccc} A & \xrightarrow{f} & Y \\ {\scriptstyle I(v)}\downarrow & \phi_{f,g}(v) \nearrow & \downarrow{\scriptstyle p} \\ B & \xrightarrow{g} & X \end{array}$$

the arrow $\phi_{f,g}(v) : B \to Y$ is a diagonal filler as drawn which makes the diagram commute. A commutative square like the above is called a *lifting problem*.

(ii) The **'horizontal' condition**: the compatibility condition that for every such $\phi_{f,g}(v)$, and every square $v' \to v$ whose image under I is given by the left-hand commutative square in the solid part of the diagram below:

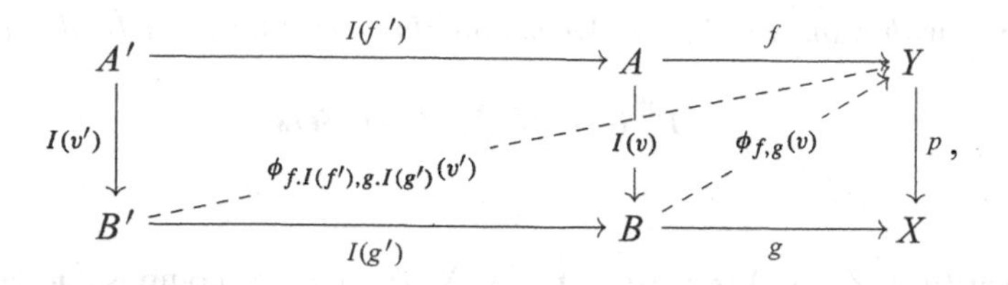

the drawn morphisms make the diagram commute, i.e.

$$\phi_{f,g}(v).I(g') = \phi_{f.I(f'),g.I(g')}(v')$$

(ii) The **'vertical' condition**: the compatibility condition that when v, w is a composable pair of vertical arrows, i.e. $\operatorname{cod} v = \operatorname{dom} w$, we have:

$$\phi_{\phi_{f,I(w).g}(v),g}(w) = \phi_{f,g}(w.v)$$

In diagrammatic notation, this means that the two ways to fill the below diagram, either in two steps or in one go, are the same:

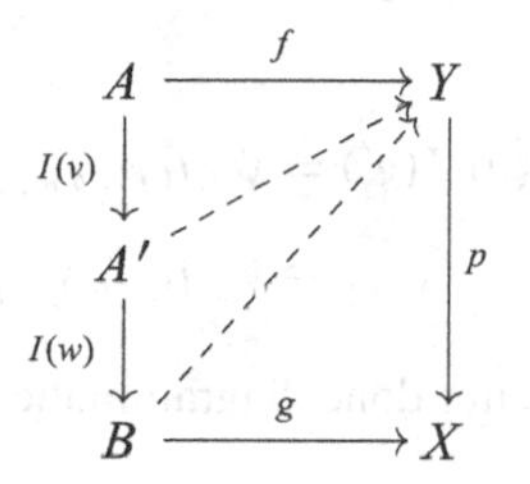

When $(u, v) : p' \to p$ is a pullback square as drawn in the diagram below, and ϕ is a right lifting structure on p with respect to I, the universal property of the pullback induces a right lifting structure $(u, v)^*\phi$ for p' with respect to I:

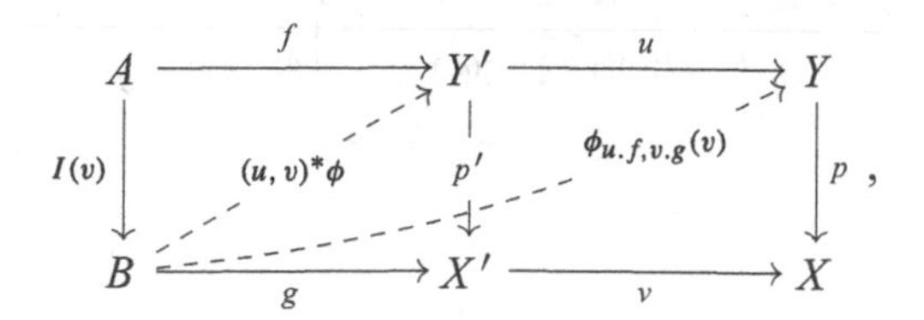

It is easy to verify that this is a right lifting structure for p'. This conclusion is summarised as follows:

Proposition 2.2 *With this pullback action, there is, for every $I : \mathbb{L} \to \mathbf{Sq}(\mathcal{E})$ as above, a fibred structure on $\mathcal{E}$ which sends each arrow to the set of right lifting* *structures with respect to* I *on it. We denote this fibred structure by the functor:*

$$I^{\pitchfork}(-) : (\mathcal{E}^{\rightarrow}_{\text{cart}})^{op} \rightarrow \mathcal{S}ets$$

□

When $(q : Z \rightarrow Y, \phi)$, $(q : Y \rightarrow X, \psi)$ are two composable arrows in $\mathcal{E}$ together with a right lifting structure with respect to I on them, there is a candidate right lifting structure on the composition $q.p$ defined by step-wise lifts:

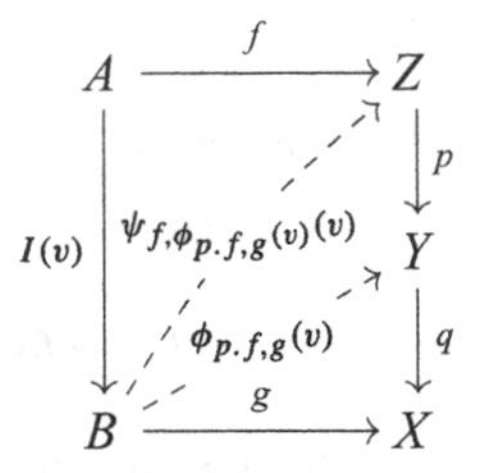

To verify that this is a right-lifting structure, we have to verify two conditions. For (ii), this is:

$$\begin{aligned}\psi_{f,\phi_{p.f,g}(v)}(v).I(g') &= \psi_{f.I(f'),\phi_{p.f,g}(v).I(g')}(v') \\ &= \psi_{f.I(f'),\phi_{p.f.I(f'),g.I(g')}(v')}(v')\end{aligned}$$

For condition (iii), this is better done diagrammatically:

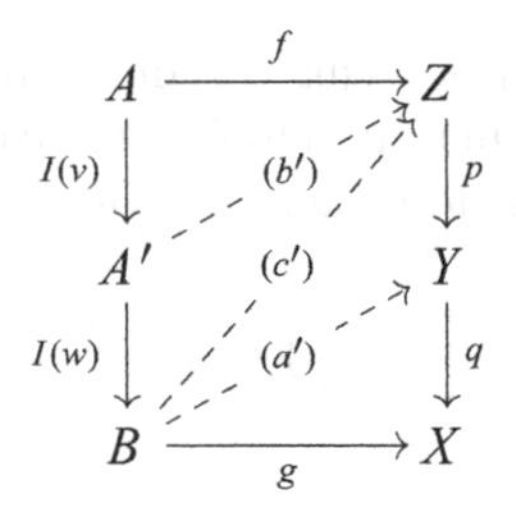

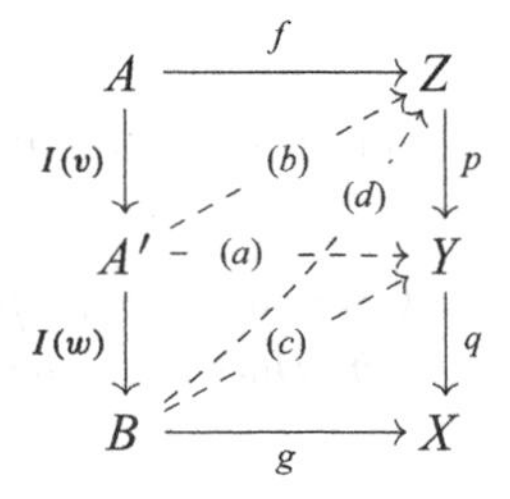

Consider the two diagrams above. On the left-hand side, a lift is obtained in two steps (a') and (c'), but we immediately note that the second step could have been done in two steps by first finding (b') and then finding the same (c'). On the right-hand side, a lift is found in four steps (a)-(d). Note that

$$(b') = \psi_{f,(a').I(w)}(v) = \psi_{f,(a)}(v) = (b)$$

And similarly:

$$(c) = \phi_{p.(b),g}(w) = \phi_{(a),g}(w) = \phi_{\phi_{p.f,I(w).g}(v),g}(w) = \phi_{p.f,g}(w.v) = (a')$$

Therefore it follows that (d) and (c') are the same. So the composed lifting structure indeed satisfies the third property above. As a consequence, the following definition is just:

Definition 2.6 Suppose $I : \mathbb{L} \to \mathbf{Sq}(\mathcal{E})$ is a double category over $\mathbf{Sq}(\mathcal{E})$. We can define a new double category $I^{\pitchfork} : \mathbb{L}^{\pitchfork} \to \mathbf{Sq}(\mathcal{E})$ as follows:

(i) Objects are objects of $\mathcal{E}$, and horizontal morphisms are morphisms in $\mathcal{E}$
(ii) A vertical morphism $Y \to X$ is a pair $(p : Y \to X, \phi)$ where ϕ is a right lifting structure for p with respect to I; that is, $\phi \in I^{\pitchfork}(p)$ in the notation of Proposition 2.2. Composition of vertical morphisms is defined as above.
(iii) A square $(p', \phi') \to (p, \phi')$ between vertical morphisms is a commutative square $p' \to p$ in $\mathcal{E}$ as on the right hand side in the diagram below, such that whenever there is a lifting problem:

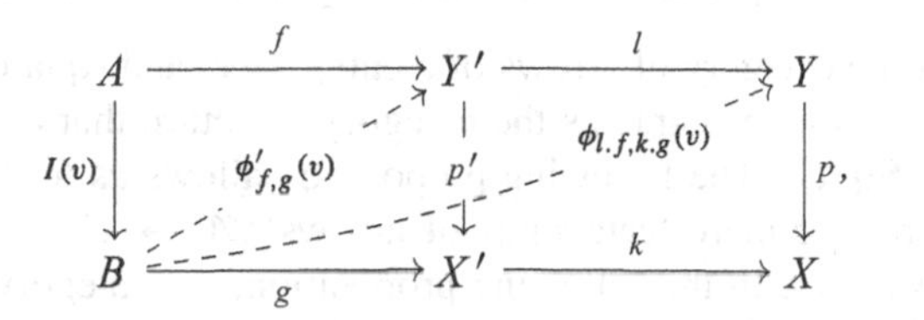

the induced diagram as drawn commutes, that is:

$$\phi_{l.f,k.g}(v) = l.\phi'_{f,g}(v).$$

Note that it needs to be checked that vertical composition of squares is compatible with the composition operation on vertical morphisms, but this follows easily from the definition of a square.

Similarly, there is a notion of *left lifting structure* for an arrow $f : A \to B$ with respect to a double category $J : \mathbb{R} \to \mathbf{Sq}(\mathcal{E})$. This consists of a family of fillers for every commutative square:

$$\begin{array}{ccc} A & \xrightarrow{f} & Y \\ {\scriptstyle f}\big\downarrow & \overset{\phi_{f,g}(v)}{\nearrow} & \big\downarrow{\scriptstyle J(v)} \\ B & \xrightarrow[g]{} & X \end{array}$$

such that three analogous conditions (i)–(iii) hold. For the sake of brevity, we will not repeat those here. In fact, the definition is completely dual to a right lifting structure, as follows. Define the *opposite* of a double category $\mathbb{A}$ to be the double category $\mathbb{A}^{\text{op}}$ with horizontal and vertical arrows as well as squares reversed, e.g. $\mathbf{Sq}(\mathcal{E})^{\text{op}} \cong \mathbf{Sq}\left(\mathcal{E}^{\text{op}}\right)$, with the obvious extension to double functors. Then a left lifting structure for an arrow f with respect to $J : \mathbb{R} \to \mathbf{Sq}(\mathcal{E})$ is the same thing as a right lifting structure for $f^{\text{op}} : B \to A$ in $\mathcal{E}^{\text{op}}$ with respect to $J^{\text{op}} : \mathbb{R}^{\text{op}} \to \mathbf{Sq}\left(\mathcal{E}^{\text{op}}\right)$. From Proposition 2.2, it follows that there is a functor

$$^{\pitchfork}J : \mathcal{E}^{\to}_{\text{cocart}} \to \mathcal{S}ets$$

where $\mathcal{E}^{\to}_{\text{cocart}}$ is the category of arrows in $\mathcal{E}$ and pushout squares between them, which sends an arrow to the set of left lifting structures on it. In other words, left lifting structure is a notion of *cofibred structure* (Definition 2.3). The following definition follows Definition 2.6:

Definition 2.7 Suppose $J : \mathbb{R} \to \mathbf{Sq}(\mathcal{E})$ is a double category over $\mathbf{Sq}(\mathcal{E})$. We can define a new double category $^{\pitchfork}J : {}^{\pitchfork}\mathbb{R} \to \mathbf{Sq}(\mathcal{E})$ analogous to Definition 2.6, but where the vertical morphisms have a left lifting structure with respect to J.

Recall that the category of arrows of a category $\mathcal{E}$ and squares between them is denoted $\mathcal{E}^{\to}$. This is the same as the category structure that exists on the vertical morphisms in $\mathbf{Sq}(\mathcal{E})$. The following proposition allows us to define left and right lifting structures for mere 'categories of arrows' $\mathcal{A} \to \mathcal{E}^{\to}$ as a special case in Definition 2.8 below. In the following proposition, the category of (small) double categories and double functors is denoted by **DBL**.

Proposition 2.3 *The functor*

$$(-)_1 : \mathbf{DBL}/\mathbf{Sq}(\mathcal{E}) \to \mathbf{Cat}/\mathcal{E}^{\to}$$

which takes a double functor $I : \mathbb{L} \to \mathbf{Sq}(\mathcal{E})$ to the category of vertical arrows and squares $I_1 : \mathbb{L}_1 \to \mathcal{E}^{\to}$ over $\mathcal{E}^{\to}$ has a fully faithful left adjoint, i.e. it is a coreflection.

Proof We will define the left adjoint

$$(-)_{\text{dbl}} : \mathbf{Cat}/\mathcal{E}^{\to} \to \mathbf{DBL}/\mathbf{Sq}(\mathcal{E})$$

as follows. Suppose $I : \mathcal{L} \to \mathcal{E}^{\to}$ is a functor. Then we can define the double category $I_{\mathrm{dbl}} : \mathcal{L}_{\mathrm{dbl}} \to \mathbf{Sq}\,(\mathcal{E})$ as:

(i) Objects of $\mathcal{L}_{\mathrm{dbl}}$ are pairs (i, v) where v is an object of $\mathcal{L}$ and $i \in \{0, 1\}$
(ii) Horizontal arrows $(i, s) : (i, u) \to (j, v)$ require that $i = j$ and are given by morphisms $s : u \to v$ in $\mathbb{L}$
(iii) The only non-identity vertical arrows are given by

$$v : (0, v) \to (1, v)$$

(iv) For every arrow $s : u \to v$ in $\mathcal{L}$, there is a square $s : 1_{(i,u)} \to 1_{(i,v)}$ for every $i \in \{0, 1\}$, and a square $s : u \to v$ between the non-identity morphisms of (ii). Vertical composition of squares is trivial, in that $s.s = s$.

Alternatively, it can be presented by the following internal category in the category of (large) categories:

$$(\mathcal{L}+\mathcal{L})+(\mathcal{L}+\mathcal{L}) \xrightarrow{[[1_{\mathcal{L}},1_{\mathcal{L}}],1_{\mathcal{L}+\mathcal{L}}]} \mathcal{L}+(\mathcal{L}+\mathcal{L}) \; \begin{array}{c} \xrightarrow{[\mathrm{inl},1_{\mathcal{L}+\mathcal{L}}]} \\ \xleftarrow{\mathrm{inr}} \\ \xrightarrow[{[\mathrm{inr},1_{\mathcal{L}+\mathcal{L}}]}]{} \end{array} \; \mathcal{L}+\mathcal{L}$$

The functor I_{dbl} then sends a square $s : u \to v$ to:

$$\begin{array}{ccc} A & \longrightarrow & B \\ {\scriptstyle I(u)}\downarrow & I(s) & \downarrow{\scriptstyle I(v)} \\ C & \longrightarrow & D \end{array}$$

It is easy to see that this construction is functorial and fully faithful. The unit is trivial, and the counit $\epsilon_I : (I_1)_{\mathrm{dbl}} \to I$ sends vertical arrows to vertical arrows, squares to squares, and sends objects $(0, v)$ to $\mathrm{dom}\, v$ and $(1, v)$ to $\mathrm{cod}\, v$, and the same for horizontal morphisms and their pointwise domain/codomain. It is easy to see that this constitutes a counit and that $(-)_1$ is a coreflection. □

Definition 2.8 Suppose $I : \mathcal{L} \to \mathcal{E}^{\to}$ is a functor. Then we can define a new functor $I^{\pitchfork} : \mathcal{L}^{\pitchfork} \to \mathcal{E}^{\to}$ as:

$$\left((I_{\mathrm{dbl}})^{\pitchfork}\right)_1 : \left((\mathcal{L}_{\mathrm{dbl}})^{\pitchfork}\right)_1 \to \mathcal{E}^{\to}.$$

Essentially, objects of $\mathcal{L}^{\pitchfork}$ are pairs $(p : Y \to X, \phi)$ where p is an arrow in $\mathcal{E}$ and ϕ is a right lifting structure with respect to the arrows in the image of I, but satisfying only the conditions (i) and (ii) since there is no non-trivial vertical composition.

For a functor $J : \mathcal{R} \to \mathcal{E}^{\to}$, there is similarly a category ${}^{\pitchfork}J : {}^{\pitchfork}\mathcal{R} \to \mathcal{E}^{\to}$ of left lifting structures satisfying only conditions (i) and (ii).

Note that in this construction, we have explicitly forgotten composition of lifting structures: even for a mere functor $I : \mathcal{L} \to \mathcal{E}^{\to}$, the category of right-lifting structures

$$(I_{\mathrm{dbl}})^{\pitchfork} : (\mathcal{L}_{\mathrm{dbl}})^{\pitchfork} \to \mathbf{Sq}\,(\mathcal{E})$$

is a double category with non-trivial vertical composition.

To conclude this section, it may be worth noting that taking double categories of left and right lifting structures are functorial constructions, and fact adjoint ones:

$$(\mathbf{DBL}/\mathbf{Sq}\,(\mathcal{E}))^{\mathrm{op}} \underset{(-)^{\pitchfork}}{\overset{{}^{\pitchfork}(-)}{\underset{\longrightarrow}{\overset{\longleftarrow}{\bot}}}} \mathbf{DBL}/\mathbf{Sq}\,(\mathcal{E})$$

for which see Proposition 18 of Bourke and Garner [23]. It follows from Proposition 2.3 that in that case also

$$(\mathbf{Cat}/\mathcal{E}^{\to})^{\mathrm{op}} \underset{(-)^{\pitchfork}}{\overset{{}^{\pitchfork}(-)}{\underset{\longrightarrow}{\overset{\longleftarrow}{\bot}}}} \mathbf{Cat}/\mathcal{E}^{\to}$$

is an adjunction. In this book, we make extensive use of the notions of left and right lifting structures in the context of algebraic weak factorisation systems, which are defined next.

2.3 Algebraic Weak Factorisation Systems

The main content of this section is a pair of equivalent definitions of an *algebraic weak factorisation system* (Definition 2.10 and Proposition 2.4). The definition of algebraic weak factorisation system (AWFS) is of foundational importance to the present work. In Chaps. 3 and 4, we demonstrate how two AWFSs can be constructed in a category $\mathcal{E}$ equipped with a *dominance* and a *Moore structure*, respectively. In the later sections, these two are combined to present the theory of effective fibrations. For the purposes of reading Chap. 3 and beyond, the definitions of AWFS referred to can be taken as a starting point.

Yet, as a starting point, these definitions involve quite a lot of assumptions on structure. The reader who would prefer to start on a basis of necessity and find out more about the ideas behind algebraic weak factorisation systems, we would encourage to study this section and the next in more depth. Indeed, as a foundation of this book, it may be worth saying a little more about the motivation behind this structure.

The main reference for the theory of AWFSs is the paper by Bourke and Garner [23], which contains the most important results off-the-shelf. Another important source is Riehl [22], who addresses some aspects in more depth, such as the subtleties around the distributive law. For the present purposes, we have already given the most important definitions in the previous section, namely that of a (left/right) lifting structure with respect to a double category. We now add to this the notion of a *functorial factorisation*.

Definition 2.9 A *functorial factorisation* for a category $\mathcal{E}$ is a section of the composition functor

$$\mathcal{E}^{\rightarrow}{}_{\mathrm{dom}} \times_{\mathrm{cod}} \mathcal{E}^{\rightarrow} \to \mathcal{E}^{\rightarrow}.$$

Spelling this out, it consists of a triple of functors $L, R : \mathcal{E}^{\rightarrow} \to \mathcal{E}^{\rightarrow}, E : \mathcal{E}^{\rightarrow} \to \mathcal{E}$, subject to two conditions. First, when f, f', h, k are morphisms in C with $f'.h = k.f$, the following diagram commutes:

$$\begin{array}{ccccc} A & \xrightarrow{Lf} & Ef & \xrightarrow{Rf} & B \\ {\scriptstyle h}\downarrow & & \downarrow{\scriptstyle E(h,k)} & & \downarrow{\scriptstyle k} \\ A' & \xrightarrow[Lf']{} & Ef' & \xrightarrow[Rf']{} & B' \end{array} \tag{2.6}$$

Second, the top and bottom composites should compose to f, f'. This decomposition yields natural transformations $\eta : 1 \Rightarrow R, \epsilon : L \Rightarrow 1$ given (at f) by the commutative squares:

$$\begin{array}{ccc} A & \xrightarrow{Lf} & Ef \\ {\scriptstyle f}\downarrow & \overset{\eta}{\Rightarrow} & \downarrow{\scriptstyle Rf} \\ B & = & B \end{array} \tag{2.7}$$

$$\begin{array}{ccc} A & = & A \\ {\scriptstyle Lf}\downarrow & \overset{\epsilon}{\Rightarrow} & \downarrow{\scriptstyle f} \\ Ef & \xrightarrow[Rf]{} & B \end{array} \tag{2.8}$$

An *algebraic weak factorisation system* (AWFS) can be informally described as a functorial factorisation into a composite of two kinds of arrows, as follows. First, there are double categories $I : \mathbb{L} \to \mathbf{Sq}(\mathcal{E})$, $J : \mathbb{R} \to \mathbf{Sq}(\mathcal{E})$ induced by each other in the following way:

$${}^{\pitchfork}J \cong I \text{ and } I^{\pitchfork} \cong J \text{ in } \mathbf{DBL}/\mathbf{Sq}(\mathcal{E}).$$

Note that also ${}^{\pitchfork}(I^{\pitchfork}) \cong I$ and $\left({}^{\pitchfork}J\right)^{\pitchfork} \cong J$. Second, it is required that the functor L of the factorisation factors through the vertical part $I_1 : \mathbb{L}_1 \to \mathcal{E}^{\to}$ (see Proposition 2.3) of I, and that R factors through the vertical part $J_1 : \mathbb{R}_1 \to \mathcal{E}^{\to}$.

Taking the above as starting point, we can work towards the formal definition of an AWFS (Definition 2.10). First, observe that if the above holds, the lifting structure ϕ on the arrows of the form Lf induces a family of maps $\delta_f = \phi_{1_{Ef},LLf}(RLf)$:

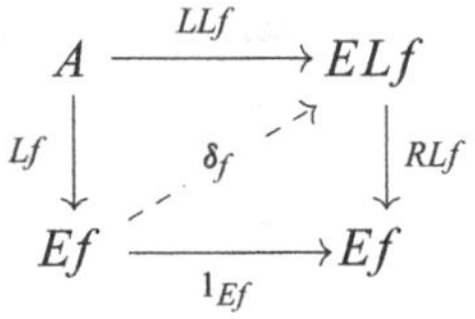

$$\begin{array}{ccc} A & \xrightarrow{LLf} & ELf \\ {\scriptstyle Lf}\downarrow & \overset{\delta_f}{\nearrow} & \downarrow{\scriptstyle RLf} \\ Ef & \xrightarrow[1_{Ef}]{} & Ef \end{array} \tag{2.9}$$

Since L factors through $\mathbb{L}_1$ and R factors through $\mathbb{R}_1$, this family defines a natural transformation:

$$\delta : L \Rightarrow LL.$$

Note that commutativity of the bottom triangle in (2.9) implies a counit law:

$$\epsilon_{Lf}.\delta_f = 1_{Lf}.$$

This counit law expresses that Lf is a coalgebra for the 'mere co-pointed endofunctor' (L, ϵ). Similarly, there is a family of maps $\mu_f : RRf \to Rf$:

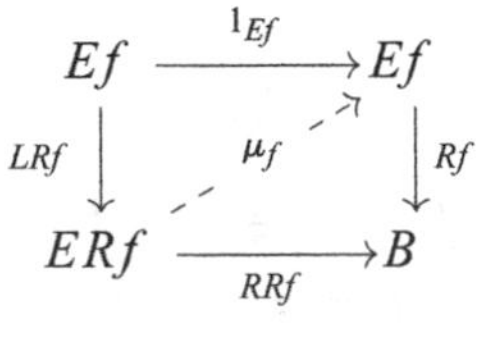

$$\begin{array}{ccc} Ef & \xrightarrow{1_{Ef}} & Ef \\ {\scriptstyle LRf}\downarrow & \overset{\mu_f}{\nearrow} & \downarrow{\scriptstyle Rf} \\ ERf & \xrightarrow[RRf]{} & B \end{array} \tag{2.10}$$

which yields a natural transformation

$$\mu : RR \Rightarrow R$$

which satisfies a unit law:

$$\mu_f.\eta_{Rf} = 1_{Rf}.$$

Again, this law just expresses that Rf is an algebra for the 'mere pointed endofunctor' (R, η). Next, there is for any vertical arrow v in the double category $\mathbb{R}$ a filler for the diagram:

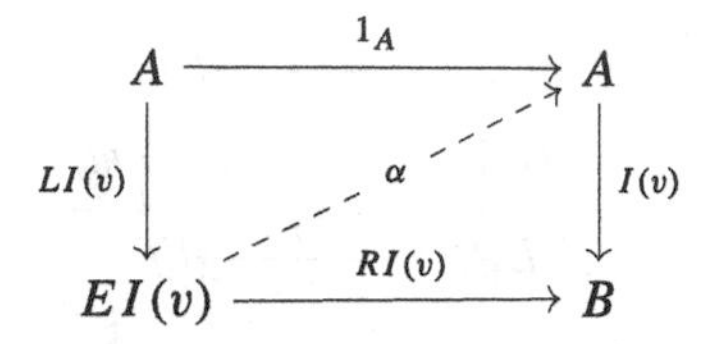

since L factors through $I_1 : \mathbb{L}_1 \to \mathcal{E}^{\to}$. Because the top triangle commutes, this filler defines an algebra structure $\alpha : RI(v) \to I(v)$ for the pointed endofunctor $R : \mathcal{E}^{\to} \to \mathcal{E}^{\to}, \eta : 1 \Rightarrow R$. Similarly, a vertical arrow u in the category $\mathbb{L}$ induces a coalgebra structure $\beta : I(u) \to LI(u)$ for the co-pointed endofunctor $L : \mathcal{E}^{\to} \to \mathcal{E}^{\to}, \epsilon : L \Rightarrow 1$.

Observe that now for any commutative square

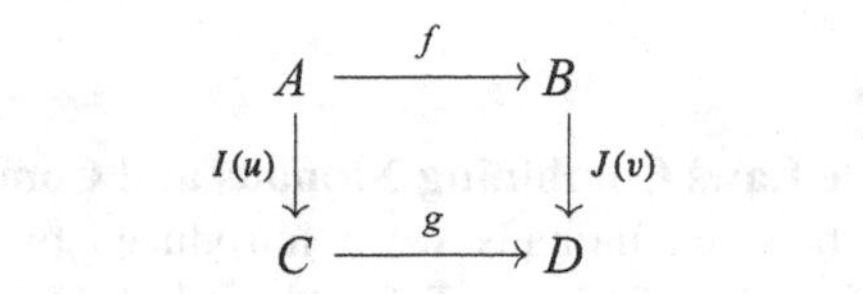

a filler can be given as a composite:

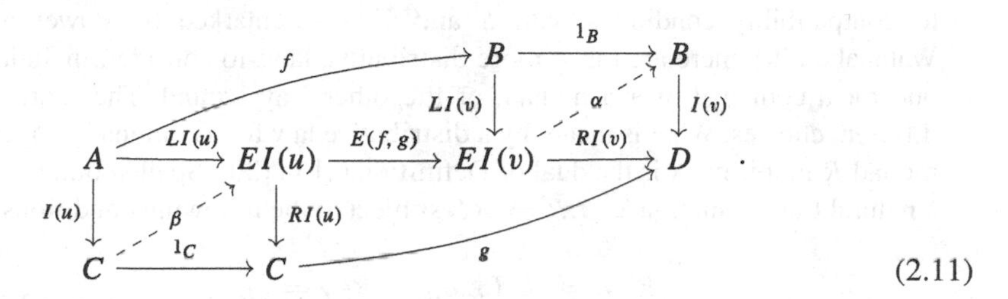

(2.11)

We now turn to the definition of an algebraic weak factorisation system (AWFS) ([23], section 2.2). A consequence of this definition is that the lifts for 'canonical' diagrams, such as (2.10) and (2.9) or more generally for any L-coalgebra and R-algebra, agree with the general shape (2.11).

Definition 2.10 Suppose (L, R, ϵ, η) is a functorial factorisation on a category $\mathcal{E}$ and $\delta : L \Rightarrow LL$, $\mu : RR \Rightarrow R$ are natural transformations. By conditions (i) and (ii) given below, we may assume that they are given by diagonals as in diagrams (2.9) and (2.10) (see [23], section 2.2, and the observations above). Then $(L, R, \delta, \mu, \epsilon, \eta)$ is an *algebraic weak factorisation system* (AWFS) when the following conditions hold:

(i) The triple (L, δ, ϵ) satisfies the conditions for a comonad structure on $\mathcal{E}^{\to}$
(ii) The triple (R, μ, η) satisfies the conditions for a monad structure on $\mathcal{E}^{\to}$
(iii) The commutative square

$$\begin{array}{ccc} Ef & \xrightarrow{\delta_f} & ELf \\ {\scriptstyle LRf}\downarrow & & \downarrow{\scriptstyle RLf} \\ ERf & \xrightarrow{\mu_f} & Ef \end{array} \tag{2.12}$$

whose diagonal is the identity $1_{Ef} : Ef \to Ef$, constitutes a *distributive law* $LR \Rightarrow RL$ of L over R. We have put some background on this in Box 2.2, but we will be mostly concerned with the alternative formulation of this condition formulated in Proposition 2.4. The significance of the distributive law for an AWFS is pointed out in Theorem 2.2 below. The fact that it constitutes a 'distributive law' in some broader sense will also become apparent from the AWFS we construct in Chap. 4.

Box 2.2 Distributive Laws Combining Monads and Comonads

Distributive laws between monads were introduced by Beck in [29] as natural transformations $\lambda : TS \Rightarrow ST$ for monads S, T on some category, subject to certain conditions relating to the monad structures of S and T. These are equivalent to a monad structure on the composition TS, subject to compatibility conditions with S and T. As remarked by Power and Watanabe [30], there are many more distributive laws to consider, including one for a comonad over a monad, or the other way around. These are all different choices. What is meant by a distributive law for a comonad L over a monad R in this book is the dual of Definition 6.1 in [30]. Spelled out, this is a natural transformation $\lambda : LR \Rightarrow RL$ subject to the following conditions:

$$\begin{gathered} R\delta.\lambda = \lambda L.L\lambda.\delta_R\,, \quad R\epsilon.\lambda = \epsilon_R \\ \mu_L.R\lambda.\lambda R = \lambda.L\mu\,, \quad \eta_L = \lambda.L\eta. \end{gathered} \tag{2.13}$$

The definition of an AWFS as given above might seem to demand an overwhelming amount of additional structure on a functorial factorisation together with the natural transformations δ, μ. The following proposition shows that in fact, the definition of an AWFS contains some redundancy and can be reduced to a couple of equational identities. For instance, the distributive law can be expressed using a single equational identity which combines δ and μ. This observation is due to Richard Garner, and since it will be quite important for our subsequent verifications, we will refer to this equation as the *Garner equation*.

Proposition 2.4 *Suppose* (L, R, ϵ, η) *is a functorial factorisation, and suppose* $\delta : L \Rightarrow LL$ *is a natural transformation over* dom : $\mathcal{E}^{\to} \to \mathcal{E}$, *and*[1] $\mu : RR \Rightarrow R$ *is a natural transformation over* cod : $\mathcal{E}^{\to} \to \mathcal{E}$. *Then the following statements are equivalent:*

(i) The triples (L, δ, ϵ), (R, μ, η) *satisfy the conditions for a comonad over* dom : $\mathcal{E}^{\to} \to \mathcal{E}$ *and a monad over* cod : $\mathcal{E}^{\to} \to \mathcal{E}$ *respectively. i.e. the following equations are satisfied:*

$$\begin{aligned} RLf.\delta_f &= 1 & \mu_f.LRf &= 1 \\ E(1, Rf).\delta_f &= 1 & \mu_f.E(Lf, 1) &= 1 \\ \delta_{Lf}.\delta_f &= E(1, \delta_f).\delta_f & \mu_f.\mu_{Rf} &= \mu_f.E(\mu_f, 1) \end{aligned} \tag{2.14}$$

And the Garner equation *holds:*

$$\delta_f.\mu_f = \mu_{Lf}.E(\delta_f, \mu_f).\delta_{Rf} \tag{2.15}$$

(ii) The diagram (2.12) *commutes, its diagonal is the identity, and constitutes a distributive law for L over R, in the sense that the Eq.* (2.13) *are satisfied.*

(iii) The tuple $(L, R, \delta, \mu, \epsilon, \eta)$ *is an AWFS.*

Proof The only thing to prove is the equivalence of (i) and (ii), since these two complement each other to an AWFS. We leave it to the reader to spell out the axioms (2.13) for λ =(2.12) and conclude that these contain precisely the counit, unit, coassociativity and associativity conditions as well as the Garner equation. □

2.4 A Double Category of Coalgebras

This section further works out the relationship between the above definition of an AWFS and the informal description of two double categories of left and right arrows which determine each other. For the purposes of this book, the most important thing to keep is that these two double categories are double categories of coalgebras and algebras with respect to the comonad L and the monad R. Chapters 3 and 4 study these coalgebras and algebras in depth for two different AWFSs.

We assume that

$$(L, R, \epsilon, \eta, \delta, \mu)$$

[1] These last two conditions mean that δ, μ are given by arrows δ_f, μ_f as above such that $\delta_f.Lf = LLf$ and $Rf.\mu_f = RRf$.

is an AWFS on a category $\mathcal{E}$ as in Definition 2.10. We would like to define a double category L - $\mathbb{C}$**oalg** whose objects are objects of $\mathcal{E}$, whose vertical arrows are coalgebras for the *comonad* (L, δ, ϵ) and whose squares are morphisms of algebras. For this to make sense, it is needed to define a vertical composition of coalgebras.

We can regard R and L as either mere (co)pointed endofunctors, or (co)monads. In both cases, (co)algebras for them can be represented as diagonal fillers for diagrams in $\mathcal{E}$. Indeed, recall that a *coalgebra* $\beta : f \to Lf$ for the mere co-pointed endofunctor L is the same thing as a filler for the square:

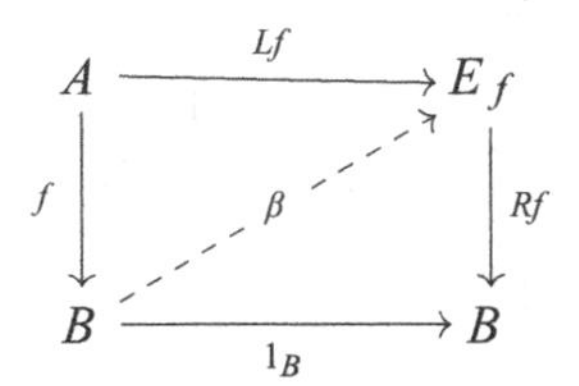

Indeed, one can check that the co-unit condition dictates that the top arrow of the square $\beta : f \to Lf$ must be the identity, hence the top triangle commutes, and that $R_f.\beta = 1_B$, hence the bottom triangle commutes. When we are interested in coalgebras for the *comonad* (L, δ, ϵ), this arrow is subject to the co-associativity condition $\delta_f.\beta = L(\beta).\beta$. This boils down to the equational identity:

$$\delta_f.\beta = E(1_A, \beta).\beta \tag{2.16}$$

Similarly, an *algebra* structure $Rf \to f$ (for either the mere pointed endofunctor or the monad) is defined entirely by an underlying arrow $\beta : E_f \to \operatorname{dom} f$. We will hence refer to algebra or coalgebra structures by their underlying map. For the comonad L and the monad R, we denote the category of (co)algebras and morphisms of (co)algebras by L - **Coalg** and R - **Alg** respectively. As we have seen in (2.11), every L-coalgebra bears a left lifting structure with respect to every R-algebra. This takes the form of a functor

$$\Phi : L\text{ - }\mathbf{Coalg} \to {}^{\pitchfork}R\text{ - }\mathbf{Alg}$$

for which we can describe the image in the form of an extra condition on lifting structures in the following proposition. This proposition is the dual of Lemma 1 in Bourke-Garner [23].

Proposition 2.5 *The functor* $\Phi : L$ - **Coalg** $\to {}^{\pitchfork}R$ - **Alg** *is injective on objects and fully faithful, and its image consists of those arrows* (i, ϕ) *with a left lifting structure* ϕ *for which the following diagrams commute:*

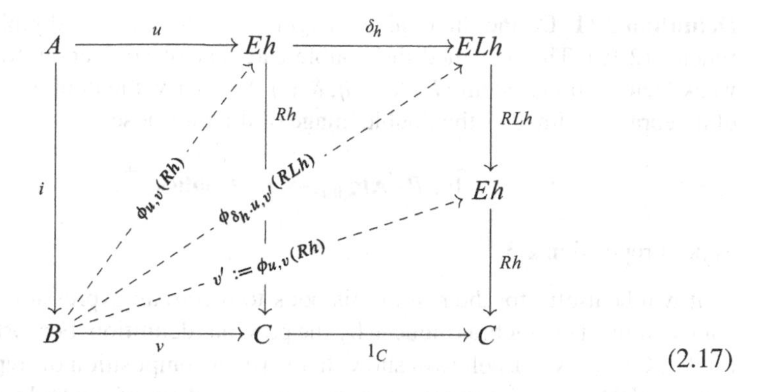

(2.17)

Proof See Lemma 1 of Bourke-Garner [23]. The proof relies on the distributive law, or more precisely, the Garner equation. □

Corollary 2.1 *Suppose we have a pushout square:*

$$\begin{array}{ccc} A & \longrightarrow & A' \\ {\scriptstyle f}\downarrow & & \downarrow{\scriptstyle f'} \\ B & \longrightarrow & B' \end{array}$$

and suppose that β is a coalgebra structure on f. Then there is a unique coalgebra structure β' on f' which makes the diagram into a morphism of coalgebras. Hence there is a cofibred structure (see the remarks above Definition 2.7) on $\mathcal{E}$:

$$L\text{-}\mathbf{coalg} : \mathcal{E}^{\rightarrow}_{cocart} \rightarrow \mathcal{S}ets \tag{2.18}$$

which associates to every morphism the set of coalgebra structures on it.

Proof There is a unique left lifting structure with respect to R-**Alg** on f' such that the square is a morphism of lifting structures. It is easy to see that this lifting structure satisfies the condition of Proposition 2.5. Therefore it also defines a coalgebra structure on f' in such a way that the diagram is a morphism of coalgebras. This defines the functor (2.18) on morphisms. □

With the construction introduced in Proposition 2.3, we can improve a bit on the above Proposition, by regarding Φ as a double functor

$$\Phi : L\text{-}\mathbf{Coalg}_{\mathrm{dbl}} \rightarrow {}^{\pitchfork}R\text{-}\mathbf{Alg}_{\mathrm{dbl}}, \tag{2.19}$$

which is injective and fully faithful on vertical morphisms. One can check that the condition on left lifting structures of Proposition 2.5 is closed under composition of left lifting structures. So we can inherit vertical composition from this category, and define a double category of coalgebras as follows:

Definition 2.11 Define the double category L - $\mathbb{C}$**oalg** as the double image of the functor (2.19). This is called the double category of coalgebras for the algebraic weak factorisation system $(L, R, \epsilon, \eta, \delta, \mu)$. Similarly, the double category R - $\mathbb{A}$**lg** of algebras is defined as the double image of the transpose

$$\widetilde{\Phi} : R\text{-}\mathbf{Alg}_{\mathrm{dbl}} \to L\text{-}\mathbf{Coalg}_{\mathrm{dbl}}{}^{\text{rth}}.$$

as per Proposition 2.3.

It will be useful for the rest of this book to record an expression for the vertical composition of coalgebras induced by the previous definition. For instance, it will be used in Corollary 4.1 below to show that vertical composition of hyperdeformation retracts (HDRs) is the same as vertical composition of coalgebras for a certain AWFS. Also, we can use it to show where the distributive law for AWFSs is actually used in the theory of Bourke and Garner [23].

So suppose that $f : A \to B$, $g : B \to C$ come with coalgebra structures $\beta : f \to Lf$, $\gamma : g \to Lg$, and write $h := g.f$. By (2.11), both f and g have the left-lifting property with respect to Rh—and indeed it turns out that their composition coalgebra structure is given by first lifting with respect to f and then g according to this recipe. Spelling out (2.11), this filler is the diagonal in the below diagram:

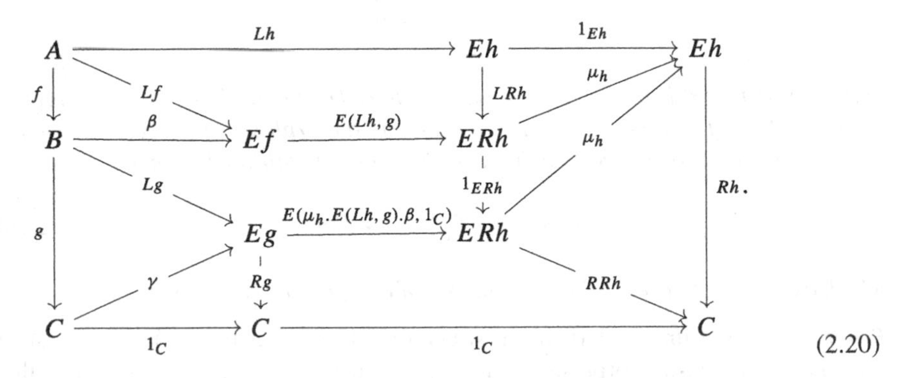

(2.20)

Hence the candidate coalgebra structure for the composite $h = g.f$ is given by

$$\begin{aligned}\kappa &:= \mu_h.E(\mu_h.E(Lh, g).\beta, 1_C).\gamma : h \to Lh \\ &= \mu_h.E(E(1_A, g).\beta, 1_C).\gamma\end{aligned} \tag{2.21}$$

where the latter condition follows from one the unit law for μ. The fact that this candidate is a coalgebra structure is a result that follows from Proposition 2.5.

Similarly, if f, g are *algebras* with algebra structures β, γ, their composition $h := g.f$ has algebra structure:

$$\beta.E(1_A, \gamma.E(f, Rh).\delta_h).\delta_h = \beta.E(1_A, \gamma.E(f, 1_C)).\delta_h \tag{2.22}$$

Lemma 2.3 *Suppose* (L, R, ϵ, η) *is a functorial factorisation and* $\delta : L \Rightarrow LL$, $\mu : RR \Rightarrow R$ *are natural transformations which make the diagrams* (2.9) *and* (2.10) *commute, i.e. with* δ_f, *every arrow* Lf *is a coalgebra for the mere co-pointed endofunctor* (L, ϵ) *and with* μ_f, *every* Rf *is an algebra for the mere pointed endofunctor* (R, η).

Then for every $h : A \to C$, *the composition* $Rh.RLh$ *has an algebra structure for the pointed endofunctor* (R, η) *given by* (2.22), *or explicitly:*

$$\kappa := \mu_{Lh}.E(1_{ELh}, \mu_h.E(RLh, 1_C)).\delta_{Rh.RLh} : E_{Rh.RLh} \to E_{Lh}$$

Further, the Garner equation (2.15) *is satisfied if and only if the following square is a morphism of algebras for the given structures:*

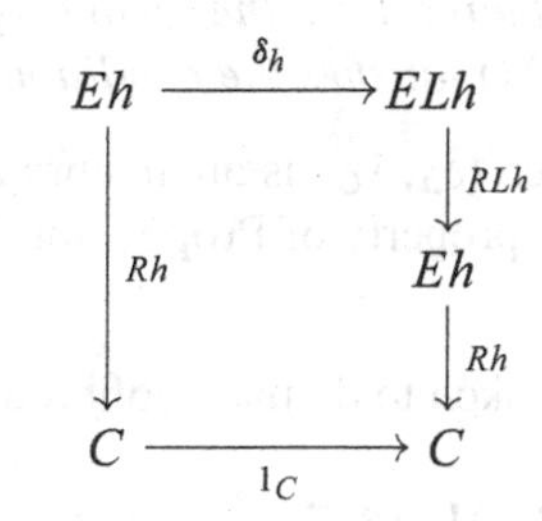

Proof The first claim follows from dualizing the preceding discussion on the composition of coalgebras for the mere co-pointed endofunctors—the filler is again the diagonal in the dual counterpart of (2.20), which is (2.22).

So we focus on the second claim. As one can readily check, this comes down to the identity:

$$\kappa.E(\delta_h, 1_C) = \delta_h.\mu_h \tag{2.23}$$

For this we have:

$$\begin{aligned}
&\kappa.E(\delta_h, 1_C) = \\
&\mu_{Lh}.E(1_{ELh}, \mu_h.E(RLh, 1_C)).\delta_{Rh.RLh}.E(\delta_h, 1_C) =^1 \\
&\mu_{Lh}.E(1_{ELh}, \mu_h.E(RLh, 1_C)).E(\delta_h, E(\delta_h, 1_C)).\delta_{Rh} = \\
&\mu_{Lh}.E(\delta_h, \mu_h.E(RLh.\delta_h, 1_C)).\delta_{Rh} = \\
&\mu_{Lh}.E(\delta_h, \mu_h).\delta_{Rh}
\end{aligned}$$

Where we used the identities

$$E(u, v).E(s, t) = E(u.s, v.t)$$

throughout, naturality of δ at $=^1$ and at the last step the identity $RLh.\delta_h = 1$. Hence (2.23) precisely states the distributive law. $\square$

From the above definitions, observe that

$$ {}^{\pitchfork}R\text{-}\mathbb{A}\mathbf{lg} \to {}^{\pitchfork}(R\text{-}\mathbf{Alg}_{\text{dbl}}) \tag{2.24} $$

is an inclusion and that Φ factors through it via the transpose of

$$ R\text{-}\mathbb{A}\mathbf{lg} \hookrightarrow (L\text{-}\mathbf{Coalg}_{\text{dbl}})^{\pitchfork}. $$

We have the following lemma:

Lemma 2.4 *Under the distributive law, the image of a vertical morphism in* ${}^{\pitchfork}R\text{-}\mathbb{A}\mathbf{lg}$ *along the functor* (2.24) *satisfies the condition of Proposition 2.5.*

Proof It is easy to see that when $(\delta_h, 1_C)$ is an algebra morphism between Rh and the composed algebra, then the property of Proposition 2.5 holds. So the statement follows from Lemma 2.3. $\square$

Since ${}^{\pitchfork}R\text{-}\mathbb{A}\mathbf{lg}$ can also be taken to define a cofibred structure

$$ {}^{\pitchfork}R\text{-}\mathbb{A}\mathbf{lg} : \mathcal{E}^{\to}_{\text{cocart}} \to \mathcal{S}ets\ , $$

Proposition 2.5 can be rewritten as a theorem on cofibred structures:

Theorem 2.2 *Suppose* $(L, R, \epsilon, \eta, \delta, \mu)$ *is an AWFS (Definition 2.10). Then:*

(i) The natural transformation between cofibred structures

$$ \varphi : L\text{-}\mathbf{coalg} \to {}^{\pitchfork}R\text{-}\mathbb{A}\mathbf{lg} $$

induced by Φ *is an isomorphism*

(ii) The natural transformation between fibred structures

$$ \widetilde{\varphi} : R\text{-}\mathbf{alg} \to L\text{-}\mathbb{C}\mathbf{oalg}^{\pitchfork} $$

induced by $\widetilde{\Phi}$ *is an isomorphism.*

Proof The two statements are dual. To prove (i), it is enough to show that that for each morphism v in $\mathcal{E}$, φ_v is an isomorphism, i.e. that coalgebra structures on v correspond precisely to left lifting structures on v with respect to $R\text{-}\mathbb{A}\mathbf{lg}$. This follows from Lemma 2.4 and Proposition 2.5. $\square$

As another consequence of the above theorem, the double functor

$$ L\text{-}\mathbf{Coalg}_{\text{dbl}} \to {}^{\pitchfork}R\text{-}\mathbb{A}\mathbf{lg} $$

through which Φ factors is surjective and full on vertical morphisms and squares apart from being injective and fully faithful on horizontal morphisms. Hence this, and similarly dual reasoning, induces equivalences of images:

$$L\text{-}\mathbb{C}\mathbf{oalg} \cong {}^{\pitchfork}R\text{-}\mathbb{A}\mathbf{lg} \text{ and } R\text{-}\mathbb{A}\mathbf{lg} \cong L\text{-}\mathbb{C}\mathbf{oalg}^{\pitchfork} \text{ (over } \mathbf{Sq}(\mathcal{E})\text{)}$$

which is the desired property of an AWFS.

In the terminology of Bourke and Garner (see Section 2.8 of [23]), the two double categories of coalgebras and algebras are *concrete* double categories over $\mathcal{E}$:

Definition 2.12 A double functor $I : \mathbb{I} \to \mathbf{Sq}(\mathcal{E})$ is called a *concrete* double category over $\mathcal{E}$ when its functor on objects (and horizontal arrows) is the identity (the objects of $\mathbb{I}$ are the objects of $\mathcal{E}$) and its functor on morphisms and squares (see Proposition 2.3)

$$I_1 : \mathbb{I}_1 \to \mathcal{E}^{\to}$$

is faithful.

To summarise the previous, we have thus found two concrete double categories over $\mathcal{E}$ which are the categories of left or right lifting structures with respect to each other. These are the double categories of coalgebras and algebras for the AWFS, respectively.

Before moving on, we address the natural question of whether there could be a different vertical composition of algebras or coalgebras than the one used above. As shown by Bourke and Garner in Proposition 4 of [23], this is not the case, in the sense that vertical composition of algebras, for a certain given monad R, completely determines an AWFS which induces that composition.

Proposition 2.6 *Suppose* $R : \mathcal{E}^{\to} \to \mathcal{E}^{\to}$ *is a monad over* cod $: \mathcal{E}^{\to} \to \mathcal{E}$. *Then there is a bijection between extensions of* R *to an AWFS and extensions of* $R\text{-}\mathbf{Alg} \to \mathcal{E}^{\to}$ *to a concrete double category over* $\mathbf{Sq}(\mathcal{E})$. *Under this bijection, the vertical composition of algebras coincides with the vertical composition induced by the AWFS.*

Proof See [23], Proposition 4. The idea is that the unit of R determines L, and δ is determined by the unique morphism of algebras $Rf \to Rf.RLf$ induced from $(LLf, 1) : f \to Rf.RLf$, since Rf has the free algebra structure on f. □

2.5 Cofibrant Generation by a Double Category

In Sect. 10.3 and further, we will compare different algebraic weak factorisation systems, or prove that they are the same (e.g., Theorem 10.2). A way to do this is to look at generating double categories. The following is from Bourke-Garner [23]:

Definition 2.13 Suppose $J : \mathbb{J} \to \mathbf{Sq}(\mathcal{E})$ is a double functor for a *small* double category $\mathbb{J}$. An AWFS is *cofibrantly generated* by $\mathbb{J}$ if $J^{\pitchfork} \cong R\text{-}\mathbb{A}\mathbf{lg}$ over $\mathbf{Sq}(\mathcal{E})$.

When $\mathbb{J}$ is large, Bourke and Garner call this property *class-cofibrantly generated.* The dual property, when ${}^{\pitchfork}I \cong L\text{-}\mathbb{C}\mathbf{oalg}$, is called *(class)-fibrantly generated.* The conclusion reached in the previous section can be summarised as:

Corollary 2.2 ([23], Proposition 20) *An AWFS is class-cofibrantly generated by its double category of coalgebras, and class-fibrantly generated by its double category of algebras.*

We refer to Bourke-Garner to results of the type that say that under appropriate conditions (i.e. $\mathcal{E}$ locally presentable), the AWFS generated by any small double category $\mathbb{J} \to \mathbf{Sq}(\mathcal{E})$ exists ([23], Proposition 23). These results rely on some kind of small object argument [31]. We will not rely on these results for our construction of an AWFS, since we work constructively from the outset. But they can be useful for boiling down a constructive theory to a classical theory for comparison.

2.6 Fibred Structure Revisited

In this preliminary chapter we have studied collections of arrows, in a category $\mathcal{E}$, equipped with structure. These collections have taken different forms, namely, that of:

1. a *notion of fibred structure* $\mathbf{fib} : \mathcal{E}^{\to}_{\text{cart}} \to \mathcal{S}ets$ (or *co-fibred* structure);
2. a *category* $I \to \mathcal{E}^{\to}$ over $\mathcal{E}^{\to}$;
3. a *concrete double category* $\mathbb{I} \to \mathbf{Sq}(\mathcal{E})$ over $\mathcal{E}$;

where we recall the definition of a *concrete* double category (Definition 2.12). We have also already seen that in many cases, the three forms are in fact three different presentations of the same thing, e.g.,:

1. The *notion of fibred structure* given by R-algebras $R\text{-}\mathbf{alg}$;
2. The *category* of R-algebras $R\text{-}\mathbf{Alg}$;
3. The *concrete double category* of R-algebras $R\text{-}\mathbb{A}\mathbf{lg}$.

The following definition would capture these three definitions into one, for the given example:

Definition 2.14 A (concrete) double category $I : \mathbb{I} \to \mathbf{Sq}(\mathcal{E})$ over $\mathbf{Sq}(\mathcal{E})$ is called *discretely fibred* when for its 1-categorical restriction (Proposition 2.3)

$$I_1 : \mathbb{I}_1 \to \mathcal{E}^{\to},$$

there is, for every object i of $\mathbb{I}_1$ and every cartesian square $\alpha : f \to I_1(i)$ in $\mathcal{E}^{\to}$, a unique object and morphism $\alpha^* : j \to i$ in $\mathbb{I}_1$, such that $I_1(\alpha^*) = \alpha$:

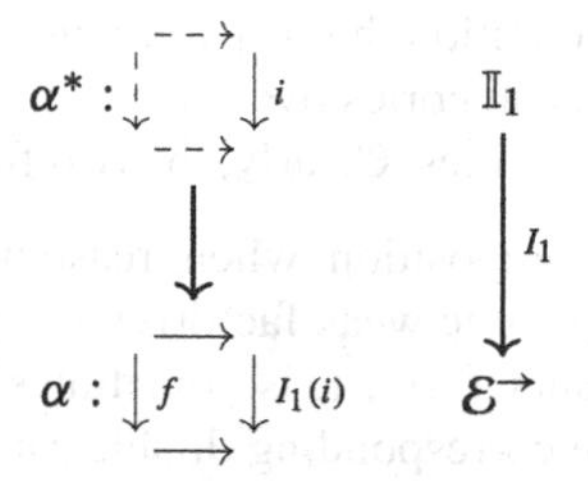

This definition has a dual definition, that of a discretely cofibred concrete double category, which comes with an induced notion of cofibred structure.

Every discretely fibred concrete double category $I : \mathbb{I} \to \mathbf{Sq}\,(\mathcal{E})$ defines a unique notion of fibred structure $\mathbf{fib}(I) : (\mathcal{E}^{\to}_{\mathrm{cart}})^{\mathrm{op}} \to \mathcal{S}ets$ on $\mathcal{E}$ with the action induced by the above property:

$$\mathbf{fib}(I)(g) = I_1^{-1}(g)$$

$$\mathbf{fib}(I)(\alpha)(i) = j.$$

Further, this operation is functorial, in that a double functor $F : \mathbb{I} \to \mathbb{J}$ between discretely fibred concrete double categories induces a natural transformation between fibred structures:

$$\mathbf{fib}(F) : \mathbf{fib}(I) \to \mathbf{fib}(J).$$

We have the following proposition:

Proposition 2.7 *Suppose* $F : \mathbb{I} \to \mathbb{J}$ *is a double functor over* $\mathbf{Sq}\,(\mathcal{E})$ *between discretely fibred concrete double categories* I *and* J *over* $\mathcal{E}$*. Then* F *is an isomorphism of double categories precisely when the following two conditions are satisfied:*

(i) F *induces a natural isomorphism of fibred structures*

$$\mathbf{fib}(F) : \mathbf{fib}(I) \cong \mathbf{fib}(J).$$

(ii) F *is full on squares, i.e., the restriction on morphisms and squares*

$$F_1 : \mathbb{I}_1 \to \mathbb{J}_1$$

is full.

Proof When F is invertible, the two conditions follow easily. For sufficiency, assume that the two conditions hold. The natural transformation $\mathbf{fib}(F)$ has a natural inverse $\mathbf{fib}(F)^{-1}$. We define the functor $G_1 : \mathbb{J}_1 \to \mathbb{I}_1$ on vertical morphisms by:

$$G_1(j) = \mathbf{fib}(F)^{-1}_{J_1(j)}(j).$$

It follows from the second condition that G_1 as defined on objects above uniquely extends to a functor between categories over $\mathcal{E}^{\rightarrow}$. Uniqueness follows here since $\mathbb{I}$ and $\mathbb{J}$ are concrete double categories. Clearly, the two functors are inverses. □

We will use the above proposition when reasoning about different double categories of arrows in an algebraic weak factorisation system, regarded as a notion of fibred structure. In the latter form, it is sometimes easier to see that they are isomorphic. To show that the corresponding double categories are isomorphic, it is only needed to show that the isomorphism is induced by a double functor which is full on squares. Note that the proposition clearly has a dual, which would hold for the dual notion of discretely cofibred concrete double categories and notions of cofibred structure.

The following result from Bourke and Garner may also be useful for the reader interested in extending double functors between double categories of algebras to morphisms between algebraic weak factorisation systems. As we have not defined AWFSs as a category above, we refer to [23] for the details. We adopt the notation from the paper, $\mathbf{AWFS}_{(\mathrm{op})\mathrm{lax}}$, for the category of algebraic weak factorisation systems and lax (resp. oplax) morphisms between them. Recall that **DBL** denotes the category of double categories and double functors between them.

Proposition 2.8 ([23], Proposition 2) *The 2-functors*

$$(-)\text{-}\mathbb{A}\mathbf{lg} : \mathbf{AWFS}_{\mathrm{lax}} \to \mathbf{DBL}^{\rightarrow}$$
$$(-)\text{-}\mathbb{C}\mathbf{oalg} : \mathbf{AWFS}_{\mathrm{oplax}} \to \mathbf{DBL}^{\rightarrow}$$

sending an AWFS to their concrete double categories of algebras, resp. coalgebras, are 2-fully faithful. □

The proposition says as much as that every double functor between double categories of (co)-algebras induces a unique (op)lax morphism of algebraic weak factorisation systems. It may be used in combination with Proposition 2.7 to conclude that a double functor between categories of algebras, which is full on squares and induces an isomorphism of notions of fibred structure, also induces an isomorphism of algebraic weak factorisation systems.

2.7 Concluding Remark on Notation

As shown in the beginning of this section, we may have to deal with the same thing throughout this book, but now as a fibred structure, then as a double category, or sometimes even just a category. To deal with this overhead, we will make sometimes make use of the notion of a discretely (co)fibred concrete double category. Yet there are also cases where we need to specifically look at this structure as a mere fibred structure or mere category. In those cases, we denote the three types of structures as **fib**, **Fib**, and $\mathbb{F}$**ib**—for (co)fibred structure, category and (concrete) double category respectively. Indeed, we have already used the notation L - **coalg**, L - **Coalg**, L - $\mathbb{C}$**oalg**, for example.

The following examples, to occur in upcoming chapters, serve as a further illustration of this notation:

- The double category of effective trivial fibrations $\mathbb{E}$**ffTrivFib**, as a category **EffTrivFib**, and as a notion of fibred structure **effTrivFib** (Chap. 3);
- The double category of hyperdeformation retractions $\mathbb{H}$**DR**, as a category **HDR**, and as a notion of cofibred structure **hdr** (Sect. 4.2);
- The double category of naive fibrations $\mathbb{N}$**Fib**, as a category **NFib**, and as notion of fibred structure **nFib** (Sect. 4.3);
- The double category of effective fibrations $\mathbb{E}$**ffFib**, as a category **EffFib**, and as a notion of fibred structure **effFib** (Sect. 6.2).

Chapter 3
An Algebraic Weak Factorisation System from a Dominance

In this chapter we show how the notion of a *dominance* gives rise to an algebraic weak factorisation system. The left class (coalgebras) of this algebraic weak factorisation system will be shown to be the class of *effective cofibrations* defined by the dominance, while the right class (algebras) is called the class of *effective trivial fibrations*. Proposition 3.1 can also be found in Bourke and Garner [23]. The rest of the chapter studies the (double) category of effective cofibrations a bit more closely and in terms of (co)fibred structure. Throughout this chapter, $\mathcal{E}$ is a category satisfying the conditions stated at the beginning of Chap. 2.

Definition 3.1 A class of monomorphisms Σ in $\mathcal{E}$ is called a *dominance* [1] on $\mathcal{E}$ if

1. every isomorphism belongs to Σ and Σ is closed under composition.
2. every pullback of a map in Σ again belongs to Σ.
3. the category Σ_{cart} of morphisms in Σ and pullback squares between them has a terminal object.

A monomorphism which is an element of Σ will be called a *effective cofibration*.

Since taking the domain of a monomorphism in Σ has a left adjoint id : $\mathcal{E} \to \Sigma_{\text{cart}}$, sending an object to the identity on it, the domain of the terminal object in Σ_{cart} is the terminal object in $\mathcal{E}$. We will denote this map by $\top : 1 \to \Sigma$. The following proposition uses that $\mathcal{E}$ is a locally cartesian closed category and so admits local exponentials. Strictly speaking, we only use that $\top : 1 \to \Sigma$ is exponentiable.

[1] As far as the authors are aware, this terminology is due to Rosolini in the context of topos theory ([32], Chap. 3). In Bourke and Garner, it is called a *stable class of monics* ([23], 4.4).

B. van den Berg, E. Faber, *Effective Kan Fibrations in Simplicial Sets*,
Lecture Notes in Mathematics 2321, https://doi.org/10.1007/978-3-031-18900-5_3

Proposition 3.1 *Let Σ be a dominance. Then the functorial factorisation given by*

$$B \xrightarrow[Lf]{} M_f = \Sigma_{a \in A} \Sigma_{\sigma \in \Sigma} B_a^\sigma \xrightarrow[Rf]{} A,$$

with $Lf(b) = (f(b), \top, \lambda x.b)$ and $Rf(a, \sigma, \tau) = a$ can be extended to an algebraic weak factorisation system.

Proof Note that Mf classifies Σ-partial maps into B over A. Let us spell out what this means. By a Σ-partial map $X \rightharpoonup B$ over A we mean a pair consisting of a subobject $m : X' \to X$ with $m \in \Sigma$ (which does not depend on the choice of representative) and a map $n : X' \to B$ making

$$\begin{array}{ccc} X' & \xrightarrow{n} & B \\ {\scriptstyle m}\downarrow & & \downarrow{\scriptstyle f} \\ X & \longrightarrow & A \end{array}$$

commute. Note that such Σ-partial maps $X \rightharpoonup B$ over A can be pulled back along arbitrary maps $Y \to X$. Saying that Mf classifies Σ-partial maps into B over A means that any such map can be obtained by pulling back the Σ-partial map $(Lf, 1_B) : M_f \rightharpoonup B$ along a unique map $X \to M_f$ over A.

For the monad structure, we need to define a map μ_f making

$$\begin{array}{ccc} M_{Rf} & \xrightarrow{\mu_f} & M_f \\ {\scriptstyle RRf}\downarrow & & \downarrow{\scriptstyle Rf} \\ A & \xrightarrow[1]{} & A \end{array}$$

commute. Maps $X \to M_{Rf}$ over A correspond to diagrams of the form

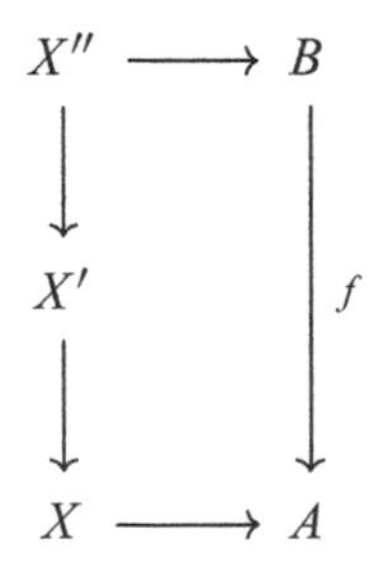

with both inclusions $X'' \to X'$ and $X' \to X$ belonging to Σ and given by pullback along $\sigma' : X' \to \Sigma$ and $\sigma : X \to \Sigma$ respectively. By considering the composition, with a corresponding map $\sigma \wedge \sigma' : X \to \Sigma$, we get a map $X \to M_f$, naturally in X, so by Yoneda we obtain a map $M_{Rf} \to M_f$ as desired. Explicitly, this map looks like

$$\mu_f : \Sigma_{a\in A}\Sigma_{\sigma\in\Sigma}\left(\Sigma_{a'\in A}\Sigma_{\sigma'\in\Sigma}B_{a'}^{\sigma'}\right)_a^{\sigma} \to \Sigma_{a\in A}\Sigma_{\sigma\in\Sigma}B_a^{\sigma}$$

$$\mu_f(a,\sigma,\chi) = \left(a, \sigma\wedge\sigma', \star\in\sigma\wedge\sigma' \mapsto (p_3.\chi.p_1(\star))(p_2(\star))\right)$$

Here, the proposition $\sigma \wedge \sigma'$ is defined as the set of pairs $(\star_1, \star_2)$ where $\star_1 \in \sigma$ and $\star_2 \in p_2.\chi(\star_1)$. One can now easily verify the unit law and associativity.

For the comonad structure, we need to define a map δ_f making

$$\begin{array}{ccc} B & \xrightarrow{1} & B \\ {\scriptstyle Lf}\downarrow & & \downarrow{\scriptstyle LLf} \\ M_f & \xrightarrow[\delta_f]{} & M_{Lf} \end{array}$$

commute. Note that

$$M_{Lf} = \sum_{(a,\sigma,\tau)\in M_f}\sum_{\sigma'\in\Sigma}(B_{(a,\sigma,\tau)})^{\sigma'}.$$

So if $((a,\sigma,\tau),\sigma',\tau') \in M_{Lf}$ and $\star \in \sigma'$, then $(a,\sigma,\tau) = (f(b), \top, \lambda x.b)$ for $b = \tau'(\star)$; hence $\star \in \sigma$ and $\tau(\star) = \tau'(\star)$. In other words,

$$M_{Lf} = \{((a \in A, \sigma \in \Sigma, \tau \in (B_a)^{\sigma}), \sigma' \in \Sigma, \tau' \in (B_a)^{\sigma}) : \sigma' \leq \sigma, \tau \restriction \sigma' = \tau'\}.$$

So we can define a map $\delta_f : M_f \to M_{Lf}$ by sending (a,σ,τ) to $((a,\sigma,\tau),\sigma,\tau)$. Counit laws and coassociativity are easily verified.

The distributive law (Garner equation) can be easily verified using the explicit notation for δ_f and μ_f:

$$\begin{aligned}
\mu_{Lf}.E(\delta_f,\mu_f).\delta_{Rf}(a,\sigma,\chi) &= \mu_{Lf}.E(\delta_f,\mu_f).((a,\sigma,\chi),\sigma,\chi)\\
&= \mu_{Lf}.(\delta_f.(a,\sigma,\chi),\sigma,\mu_f.\chi)\\
&= (\mu_f.(a,\sigma,\chi), \sigma\wedge\sigma', \star\in\sigma\wedge\sigma' \mapsto (p_3.\delta_f.\chi.p_1(\star)))\\
&= (\mu_f.(a,\sigma,\chi), \sigma\wedge\sigma', \star\in\sigma\wedge\sigma' \mapsto (p_3.\chi.p_1(\star)))\\
&= \delta_f.\mu_f(a,\sigma,\chi)
\end{aligned}$$

□

Proposition 3.2 (Effective Cofibrations are Precisely Coalgebras) *A coalgebra structure for $f : B \to A$ is unique, and it exists if and only if f belongs to Σ.*

Hence, there is a cofibred structure

$$\sigma : \mathcal{E}_{\text{cocart}} \to \mathcal{S}ets \tag{3.1}$$

where $\sigma(f)$ contains a single element when $f \in \Sigma$, and is empty otherwise. Moreover, there is an isomorphism of cofibred structures between σ and the cofibred structure of coalgebras.

Proof We show that every $f : B \to A$ can be equipped with a coalgebra structure for the copointed endofunctor M if and only if it belongs to Σ, and that the coalgebra structure is indeed unique and always satisfies the coassociativity condition. From this, it is easy to derive an isomorphism of cofibred structures in light of (3.1).

Suppose $\gamma : A \to M_f$ is a map exhibiting f as a coalgebra for the copointed endofunctor M. In other words, we have $Rf.\gamma = 1_A$ and γ makes

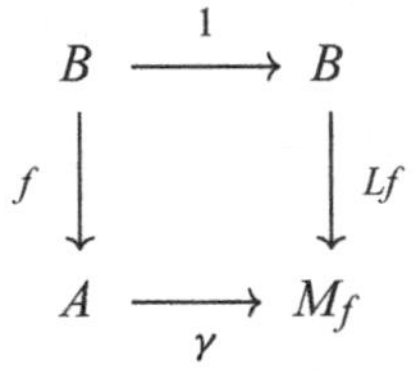

$$\begin{array}{ccc} B & \xrightarrow{1} & B \\ {\scriptstyle f}\downarrow & & \downarrow{\scriptstyle Lf} \\ A & \xrightarrow[\gamma]{} & M_f \end{array}$$

commute. These data correspond to a Σ-partial map $A \rightharpoonup B$:

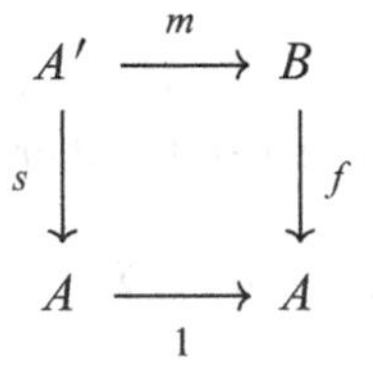

$$\begin{array}{ccc} A' & \xrightarrow{m} & B \\ {\scriptstyle s}\downarrow & & \downarrow{\scriptstyle f} \\ A & \xrightarrow[1]{} & A \end{array}$$

where $s \in \Sigma$. Further, s fits in a pullback square, namely the middle square of the following diagram:

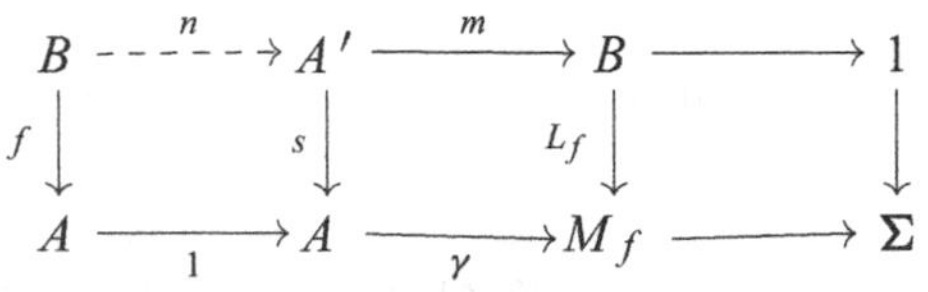

$$\begin{array}{ccccccc} B & \overset{n}{\dashrightarrow} & A' & \xrightarrow{m} & B & \longrightarrow & 1 \\ {\scriptstyle f}\downarrow & & {\scriptstyle s}\downarrow & & {\scriptstyle L_f}\downarrow & & \downarrow \\ A & \xrightarrow[1]{} & A & \xrightarrow[\gamma]{} & M_f & \longrightarrow & \Sigma \end{array}$$

whence we find $n : B \to A'$ such that $m.n = 1$. Because $s.n.m = f.m = s$ and s is monic, we also have $n.m = 1_{A'}$. So A' and B are isomorphic over A. It follows $f \in \Sigma$ and that γ classifies the map $(f, 1_B)$. From this it is clear that γ

must be unique whenever it exists and that it will always satisfy the coassociativity condition. It also follows, incidentally, that the square at the beginning of the proof is a pullback.

Conversely, if $f \in \Sigma$ we can choose $s = f$ and $m = 1$ and this gives us the coalgebra structure for the copointed endofunctor we want. The second part of the proposition follows immediately. □

Lastly, we briefly stop at algebras for the monad. From Theorem 2.2, we know that the fibred structure of algebras is isomorphic to the fibred structure of right lifting structures with respect to the *double* category of coalgebras. It remains to characterize this category. To that end, we use the following:

Lemma 3.1 *If $f : B \to A$ and $f' : B' \to A'$ are coalgebras, then a pair of maps $(m : B' \to B, n : A' \to A)$ making*

$$\begin{array}{ccc} B' & \xrightarrow{m} & B \\ {\scriptstyle f'}\downarrow & & \downarrow{\scriptstyle f} \\ A' & \xrightarrow[n]{} & A \end{array}$$

commute is a morphism of coalgebras if and only if the square is a pullback.

Proof It is not hard to check that such a pullback square constitutes a morphism of coalgebras.

For the converse, let us first make the following observation. Suppose $\gamma : A \to M_f$ is a coalgebra structure on f. Then γ fits into a diagram of the form

$$\begin{array}{ccccc} B & \xrightarrow{1} & B & \longrightarrow & 1 \\ {\scriptstyle f}\downarrow & & \downarrow{\scriptstyle Lf} & & \downarrow{\scriptstyle \top} \\ A & \xrightarrow[\gamma]{} & M_f & \longrightarrow & \Sigma \end{array}$$

where $M_f \to \Sigma$ is the obvious projection. Note that the right hand square is always a pullback and that the left hand square is as well, as we saw in the previous proof. So the outer rectangle is a pullback.

So if

$$\begin{array}{ccc} B' & \xrightarrow{m} & B \\ {\scriptstyle f'}\downarrow & & \downarrow{\scriptstyle f} \\ A' & \xrightarrow[n]{} & A \end{array}$$

is a morphism of coalgebras, then this fits into a commutative diagram of the form

$$\begin{array}{ccccc} B' & \xrightarrow{m} & B & \longrightarrow & 1 \\ {\scriptstyle f'}\big\downarrow & & \big\downarrow{\scriptstyle f} & & \big\downarrow{\scriptstyle \top} \\ A' & \xrightarrow[n]{} & A & \longrightarrow & \Sigma \end{array}$$

in which the right hand square and the outer rectangle are pullbacks. Therefore the left hand square is a pullback as well. □

Hence we can deduce:

Corollary 3.1 *Let $\mathbf{\Sigma}$ be the double category whose horizontal arrows are arbitrary arrows from $\mathcal{E}$, whose vertical arrows are maps from Σ and whose squares are pullbacks squares. Then:*

(i) There is an isomorphism between the following notions of fibred structure:

- *Algebras for the AWFS induced by Σ;*
- *Right lifting structures with respect to $\mathbf{\Sigma}$.*

(ii) There is a functor

$$R\text{-}\mathbf{Alg} \to \mathbf{\Sigma}^{\pitchfork}$$

given on objects by (i) which is fully faithful.

(iii) There is an equivalence of double categories

$$R\text{-}\mathbb{A}\mathbf{lg} \cong \mathbf{\Sigma}^{\pitchfork\!\!\pitchfork}$$

whose vertical restriction is prescribed by (ii).

Proof Because $\mathbf{\Sigma}$ is, essentially, the double category of coalgebras for the comonad. □

Definition 3.2 The algebras or right lifting structure of Corollary 3.1 are called *effective trivial fibrations*. The discretely fibred concrete double category of effective trivial fibrations is denoted $\mathbb{E}\mathbf{ffTrivFib}$, but we may also refer to it as category (**EffTrivFib**) or notion of fibred structure (**effTrivFib**).

We will return to effective trivial fibrations in Sect. 6.2.1, where we show that they are also effective fibrations (introduced in Chap. 6). Effective trivial fibrations are important for developing a theory of effective Kan fibrations in simplicial sets. For example, they allow us to relate our notion of effective Kan fibration to other notions of fibration (see Corollary 11.2 and below).

Chapter 4
An Algebraic Weak Factorisation System from a Moore Structure

In this chapter we construct an algebraic weak factorisation system on a *category with Moore structure* $\mathcal{E}$. We will then study its coalgebras and algebras more closely. We show that the structure of a coalgebra is equivalent to the structure of a *hyperdeformation retract* (HDR). Algebras, on the other hand, will be identified as *naive fibrations*.

Categories with Moore structure are a modification of the *path object categories* introduced by Van den Berg-Garner [19]. In that paper, a *cloven weak factorisation system* (see Section 3.2 loc. cit.) is constructed for an arbitrary path object category. Here, we use a new method to show that it is also possible to construct an *algebraic* weak factorisation system. To that end, we need to modify the axioms of a Moore structure relative to the original definition—but note that ours is neither weaker nor stronger.

Definition 4.1 [Moore Structure] Let $\mathcal{E}$ be a category with finite limits. A *Moore structure* for $\mathcal{E}$ is a pullback-preserving endofunctor $M : \mathcal{E} \to \mathcal{E}$ together with:

(i) Natural transformations μ, r, s, t with domain and codomains defined by the diagram:

$$MX \times_{(t,s)} MX \xrightarrow{\ \mu_X\ } MX \underset{s_X}{\overset{t_X}{\rightrightarrows}} X \quad (r_X : X \to MX),$$

Further, this data gives every object X in $\mathcal{E}$ the structure of an internal category with arrow object MX. The object MX is to be interpreted as the object of *paths* in X with *source* and *target* maps s, t, as well as a trivial path r. The object $M1$ is the object of *path lengths* and hence $M! : MX \to M1$ takes a path to its length.

(ii) A natural transformation $\Gamma : M \Rightarrow MM$, which turns (M, s, Γ) into a comonad. The idea is that Γ_X sends a path γ in MX to a 'path of paths' in

B. van den Berg, E. Faber, *Effective Kan Fibrations in Simplicial Sets*,
Lecture Notes in Mathematics 2321, https://doi.org/10.1007/978-3-031-18900-5_4

MMX, of the same length as γ, which contracts γ to the trivial path at $t_X(\gamma)$ (see Box 4.1).

(iii) A natural transformation α given by a family $\alpha_X : X \times M1 \to MX$ called *strength* (but see (2) below and Definition A.1). This map sends a point and a length to the constant path of that length.

These conditions are subject to further conditions on interactions between them. The full definition can be found in the Appendix, Definition A.1. There we make each of the conditions (i)–(iii) more precise and we present the further conditions on interaction between these elements. The most important one is the *distributive law*, which we also spell out in (2) below.

Box 4.1 The Path Contraction Γ

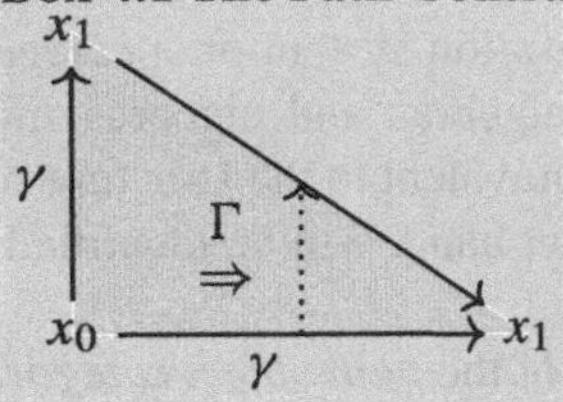

The natural transformation Γ applied to a path γ can be viewed as a contraction of γ to the trivial path on its end point. The contraction has the same length as the path itself. Each transversal 'fibre' of consists of a final segment of γ (illustrated by the dotted arrow), namely that which remains after one has contracted along the segment before. This intuition is contained in the distributive law (see Definition A.1 (5) in the Appendix). The top path is obtained by taking the pointwise target of $\Gamma(\gamma)$, which is the constant path of the same length as γ.

The differences between Definition 4.1 (and Definition A.1) and the one in [19] are the following:

1. The coassociativity condition:

$$M\Gamma.\Gamma = \Gamma M.\Gamma : MX \to MMX.$$

which turns Γ into a comonad;

2. The distributive law:

$$\Gamma.\mu = \mu.(M\mu.\nu.(\Gamma.p_1, \alpha.(p_2, M!.p_1)), \Gamma.p_2) : MX \times_X MX \to MMX.$$

where we have defined the natural transformation $\alpha : X \times M1 \to X$ by $\alpha_X := \theta_{MX}.\alpha_{MX,1}$ with $\theta, \alpha_{-,-}$ as in Remark A.2 in the Appendix. A diagrammatic illustration of this condition can be found in Eq. (A.1) in the same place.

3. The 'twist map' $\tau : M \Rightarrow M$ is dropped, only to return in Chap. 5 as *symmetric* Moore structure.

In addition, we will also introduce a notion of *symmetric* and *two-sided* Moore structure in Chap. 5. In Example A.1 in the Appendix we have given several examples of categories with (symmetric) Moore structure. In Part II of this book, we will show that the category of simplicial sets is a category with Moore structure, for the simplicial Moore path functor defined in [19].

Remark 4.1 (Path Contraction vs. Connections) The 'path contraction' Γ of Definition 4.1 can be compared to the notion of *connection* in the category of Cubical Sets [9], and more abstractly in categories with a functorial cylinder [8]. In Appendix B, we have worked out that indeed, connections in cubical sets can be used to define a path contraction Γ for a standard choice of path object. However, since all paths in this object have the same length, it is not possible to define a suitable notion of path composition for this object.

4.1 Defining the Algebraic Weak Factorisation System

This section shows how every Moore structure bears an algebraic weak factorisation system. A similar result, in absence of distributive laws, can also be found in North ([33], Theorem 3.28). Here, we improve on this by showing that our distributive law for Moore structures implies the distributive law for an AWFS. In the short sections that follow, we address each one of the requirements of Proposition 2.4 (i).

4.1.1 Functorial Factorisation

First of all, if $f : A \to B$ is a morphism, we can factor it as

$$A \xrightarrow{Lf} Ef \xrightarrow{Rf} B$$

by putting $Ef = MB \times_B A$ (pullback of t and f), $Lf = (r.f, 1)$ and $Rf = s.p_1$. In the obvious way E, L and R extend to functors, and the whole factorisation is readily seen to be functorial.

4.1.2 The Comonad

The comultiplication δ_f needs to be a mapping filling:

$$\begin{array}{ccc} A & \xrightarrow{\quad 1 \quad} & A \\ {\scriptstyle Lf=(r.f,1)}\downarrow & & \downarrow{\scriptstyle LLf=(r.(r.f,1),1)} \\ MB \times_B A & \longrightarrow & M(MB \times_B A) \times_{MB \times A} A \end{array}$$

where the object in the lower right-hand corner is the pullback of $t_{MB \times_B A}$ and Lf. Note that there is a mediating natural isomorphism

$$(MMB \times_{MB} MA) \times_{MB \times_B A} A \xrightarrow{(\nu,1)} M(MB \times_B A) \times_{MB \times_B A} A$$

where ν is the natural isomorphism induced by pullback-preservation of M. Hence we can put:

$$\delta_f = (\nu.(\Gamma.p_1, \alpha.(p_2, M!.p_1)), p_2). \tag{4.1}$$

Curiously, δ_f in no way refers to f. We can check it is well-defined. For the part mapping into $MMB \times_{MB} MA$, we have:

$$\begin{aligned} Mt.\Gamma.p_1 =& \alpha.(t, M!).p_1 \\ =& \alpha.(t.p_1, M!.p_1) \\ =& \alpha.(f.p_2, M!.p_1) \\ =& Mf.\alpha.(p_2, M!.p_1). \end{aligned}$$

Further, $t.\Gamma.p_1 = r.t.p_1 = r.f.p_2$. In addition, the square commutes, as one easily checks.

From here it remains to check the comonad laws given in Proposition 2.4. First of all:

$$\begin{aligned} s.p_1.\delta_f =& (s.\Gamma.p_1, s.\alpha.(p_2, M!.p_1)) \\ =& (p_1, p_2) \\ =& 1 \end{aligned}$$

and

$$\begin{aligned} (M(s.p_1).p_1, p_2).\delta_f =& (Ms.\Gamma.p_1, p_2) \\ =& (p_1, p_2) \\ =& 1, \end{aligned}$$

and hence the counit laws are satisfied. Second, we check coassociativity:

$$
\begin{aligned}
E(1, \delta_f) =& (M\delta_f.p_1, p_2).\delta_f = \\
&= (\nu.(\nu.(M\Gamma.Mp_1.p_1, M\alpha.\nu.(Mp_2.p_1, MM!.Mp_1.p_1)), Mp_2.p_1), p_2).\delta_f \\
&= (\nu.(\nu.(M\Gamma.\Gamma.p_1, M\alpha.(\alpha.(p_2, M!.p_1), MM!.\Gamma.p_1)), \\
&\qquad M\alpha.(p_2, M!.p_1)), p_2) \\
&=^1 (\nu.(\nu.(\Gamma.\Gamma.p_1, \Gamma.\alpha.(p_2, M!.p_1)), \alpha.(p_2, M!.p_1)), p_2) \\
&=^2 (\nu.(\Gamma.p_1, \alpha.(p_2, M!.p_1)), p_2).\delta_f \\
&= \delta_{L_f}.\delta_f
\end{aligned}
$$

where we have used the strength of Γ with respect to α at $=^1$ and the identity

$$M!.\nu.(\Gamma.p_1, \alpha.(p_2, M!.p_1)) = M!.p_1$$

at $=^2$.

4.1.3 The Monad

The multiplication is given by:

$$
\begin{array}{ccc}
MB \times_{(t,s.p_1)} (MB \times_B A) & \xrightarrow{\mu_f} & MB \times_B A \\
\downarrow{\scriptstyle s.p_1} & & \downarrow{\scriptstyle s.p_1} \\
B & \xrightarrow{1} & B
\end{array}
\tag{4.2}
$$

with $\mu_f = (\mu.(p_1, p_1.p_2), p_2.p_2)$, reminding the reader that we write the arguments to μ in sequential order rather than order of categorical composition (μ is to be thought of as path concatenation). Also note that again, the definition of the multiplication formula in no way refers to f. The monad laws are easy given that MB is an internal category.

4.1.4 The Distributive Law

We should check the Garner equation (2.15), given by the identity

$$\delta_f.\mu_f = \mu_{Lf}.E(\delta_f, \mu_f).\delta_{Rf} \tag{4.3}$$

for every f. We unfold (4.3) for the definitions presented above. We do so by (as before) explicitly by writing ν for the (family of) mediating isomorphisms:

$$\nu : MA \times_{MC} MB \to M(A \times_C B) \tag{4.4}$$

For this, the expression for the left-hand side of (4.3) amounts to:

$$\begin{aligned}
&\delta_f.\mu_f &&= \\
&(\nu.(\Gamma.p_1, \alpha.(p_2, M!.p_1)), p_2).(\mu.(p_1, p_1.p_2), p_2.p_2) &&= \\
&(\nu.(\Gamma.\mu.(p_1, p_1.p_2), \alpha.(p_2.p_2, M!.\mu.(p_1.p_2, p_1))), p_2.p_2)
\end{aligned}$$

The right-hand side amounts to:

$$\begin{aligned}
&\mu_{Lf}.E(\delta_f, \mu_f).\delta_{Rf} &&= \\
&(\mu.(p_1, p_1.p_2), p_2.p_2). \\
&(M(\mu.(p_1, p_1.p_2), p_2.p_2).p_1, (\nu.(\Gamma.p_1, \alpha.(p_2, M!.p_1)), p_2).p_2). \\
&(\nu.(\Gamma.p_1, \alpha.(p_2, M!.p_1)), p_2) &&= \\
&(\mu.(p_1, p_1.p_2), p_2.p_2). \\
&(M(\mu.(p_1, p_1.p_2), p_2.p_2).\nu.(\Gamma.p_1, \alpha.(p_2, M!.p_1)), (\nu.(\Gamma.p_1, \alpha.(p_2, M!.p_1)), p_2).p_2) &&= \\
&(\mu.(M(\mu.(p_1, p_1.p_2), p_2.p_2).\nu.(\Gamma.p_1, \alpha.(p_2, M!.p_1)), \\
&\quad p_1.(\nu.(\Gamma.p_1, \alpha.(p_2, M!.p_1)), p_2).p_2), p_2.p_2)
\end{aligned}$$

Since the second components of the two expressions are the same, we can focus on the first component. We proceed by reducing this first component for the right-hand side as follows:

$$\begin{aligned}
&\mu.(M(\mu.(p_1, p_1.p_2), p_2.p_2).\nu.(\Gamma.p_1, \alpha.(p_2, M!.p_1)), \\
&\qquad p_1.(\nu.(\Gamma.p_1, \alpha.(p_2, M!.p_1)), p_2).p_2) = \\
&\mu.(M(\mu.(p_1, p_1.p_2), p_2.p_2).\nu.(\Gamma.p_1, \alpha.(p_2, M!.p_1)), \\
&\qquad \nu.(\Gamma.p_1, \alpha.(p_2, M!.p_1)).p_2) =^1 \\
&\mu.(\nu.(M\mu.\nu.(p_1, M(p_1).p_2), M(p_2).p_2).(\Gamma.p_1, \alpha.(p_2, M!.p_1)), \\
&\qquad \nu.(\Gamma.p_1, \alpha.(p_2, M!.p_1)).p_2) = \\
&\mu.(\nu.(M\mu.\nu.(\Gamma.p_1, M(p_1).\alpha.(p_2, M!.p_1)), M(p_2).\alpha.(p_2, M!.p_1)), \\
&\qquad \nu.(\Gamma.p_1, \alpha.(p_2, M!.p_1)).p_2) =^2 \\
&\mu.(\nu.(M\mu.\nu.(\Gamma.p_1, \alpha.(p_1.p_2, M!.p_1)), \alpha.(p_2.p_2, M!.p_1)),
\end{aligned}$$

$$
\begin{aligned}
&\nu.(\Gamma.p_1.p_2, \alpha.(p_2.p_2, M!.p_1.p_2))) =^3 \\
&\nu.(\mu.(M\mu.\nu.(\Gamma.p_1, \alpha.(p_1.p_2, M!.p_1)), \Gamma.p_1.p_2), \\
&\quad \mu.(\alpha.(p_2.p_2, M!.p_1), \alpha.(p_2.p_2, M!.p_1.p_2))) = \\
&\nu.(\Gamma.\mu.(p_1, p_1.p_2), \alpha.(p_2.p_2, M!.\mu.(p_1.p_2, p_1)))
\end{aligned}
$$

Here $=^1$ uses the identity

$$\nu.(M\mu.(p_1, Mp_1.p_2), Mp_2.p_2) = M(\mu.(p_1, p_1.p_2), p_2.p_2).\nu$$

which is just naturality of ν. At $=^2$ we have twice used naturality of α and absorbed a projection term in the right term of the composite. At that point, we are left with a nice expression of the form:

$$\mu.\nu \times \nu : (MA \times_{MB} MMB) \times_{A \times_B MB} (MA \times_{MB} MMB) \to (MA \times_{MB} MMB)$$

Naturality of μ and ν implies that this map can be rewritten as

$$\nu.(\mu.(p_1.p_1, p_1.p_2), \mu.(p_2.p_1, p_2.p_2)) \tag{4.5}$$

which we have done at $=^3$. In the resulting expression, we recognize in the right component the composition of two constant paths, which is the constant path on the composition, and in the left component the distributed term of the distributive law between μ and Γ, which yields the desired equality.

We can summarise the result of this section as follows:

Proposition 4.1 *Suppose $\mathcal{E}$ is a category with Moore structure. Then the functorial factorisation (L, R, ϵ, η) given by*

$$A \xrightarrow{Lf := (r.f,\ 1)} MB \times_B A \xrightarrow{Rf := s.p_1} B$$

together with the natural transformations $\delta : L \Rightarrow LL$ defined by (4.1) *and $\mu : RR \Rightarrow R$ defined by* (4.2) *constitutes an algebraic weak factorisation system (AWFS) on $\mathcal{E}$ in the sense of Definition 2.10.* □

4.2 Hyperdeformation Retracts

We start with defining *hyperdeformation retractions* and *hyperdeformation retracts.*

Definition 4.2 Let $i : A \to B$ be a map. To equip i with a *hyperdeformation retraction* means specifying a map $j : B \to A$ and a homotopy $H : B \to MB$ such

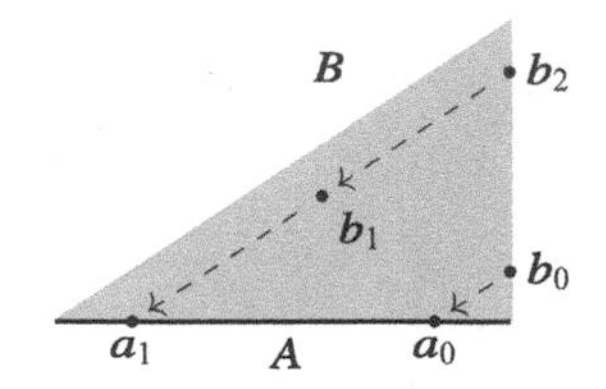

Fig. 4.1 A *hyperdeformation retract* (HDR) $i : A \rightarrow B$. If an element b_2 is retracted along a path to an element a_1, then any intermediate element b_1 is retracted along the remainder of that path

that the following hold:

$$j.i = 1_A, \quad s.H = 1_B, \quad t.H = i.j, \quad MH.H = \Gamma.H.$$

If such a structure can be specified for i, we will call it a *hyperdeformation retract*, or *HDR*.

The intuition behind a hyperdeformation retract $i : A \rightarrow B$ is as follows. First, the family of paths $H : B \rightarrow MB$ forms a *deformation retraction* of B onto A, contracting every element in B to an element in A along the path $H(b)$. Second, every other element in B lying on this path $H(b)$ retracts along the remaining segment of $H(b)$ defined by it. This is expressed by the last condition involving Γ. See Fig. 4.1 for an illustration of this condition.

The notion of a hyperdeformation retraction is a strengthening of the notion of a *strong* deformation retraction (see Van den Berg-Garner [19], Definition 6.1.3). This notion is also used by Gambino and Sattler in the context of (algebraic) weak factorisation systems (See [8], Section 3 and [26]) and by Cisinski in the context of homotopical algebra (Cisinski [34], Section 2.4). A deformation retraction is *strong* if (in the notation of Definition 4.2) $H.i = r.i$, i.e., the retraction restricted to A is trivial. Every hyperdeformation retraction is automatically strong, as is formally shown in the proof of Proposition 4.3 below.

Remark 4.2 The maps t and Γ equip any $r : X \rightarrow MX$ with the structure of a hyperdeformation retraction.

Proposition 4.2 *The function which associates to every* $i : A \rightarrow B$ *the set*

$$\{(H, j) \mid (i, j, H) \text{ is an HDR}\}$$

can be extended to a presheaf on the category of arrows of $\mathcal{E}$ *and cocartesian (pushout) squares:*

$$\mathbf{hdr} : \mathcal{E}^{\rightarrow}_{\text{cocart}} \rightarrow \mathcal{S}ets.$$

So HDRs form a cofibred structure on $\mathcal{E}$.

Proof It is sufficient to show that HDRs are closed under pushouts in a way compatible with composition of pushout squares. The reader is invited to do this as an exercise, as it will also be clear from the proof of Proposition 4.3. □

Proposition 4.3 *The following notions of cofibred structure are isomorphic:*

- *Having a coalgebra structure with respect to* $L = (r.f, 1)$,
- *Having the structure of an HDR.*

Proof Suppose $i : A \to B$ is an arrow in $\mathcal{E}$. The map $i : A \to B$ carries a coalgebra structure if there is a map (H, j) making

$$\begin{array}{ccc} A & \xrightarrow{1} & A \\ {\scriptstyle i}\downarrow & & \downarrow{\scriptstyle (1,r.i)} \\ B & \xrightarrow[(H,j)]{} & MB \times_B A \end{array}$$

commute (which means $i.j = t.H$, $j.i = 1$, $H.i = r.i$) and such that $s.p_1.(H, j) = 1$ (that is, $s.H = 1$) and $\delta.(H, j) = (p_1.M(H, j), p_2).(H, j)$. The latter condition means that

$$\delta.(H, j) = (\nu.(\Gamma.p_1, \alpha.(p_2, M!.p_1)), p_2).(H, j) = (\nu.(\Gamma.H, \alpha.(j, M!.H)), j)$$

should equal

$$(M(H, j).p_1, p_2).(H, j) = (\nu.(MH.H, Mj.H), j)$$

(using naturality of ν). In other words, that $\alpha.(j, M!.H) = Mj.H$ and $\Gamma.H = MH.H$. To summarise, a coalgebra structure is a hyperdeformation retraction satisfying, additionally, $H.i = r.i$ and $Mj.H = \alpha.(j, M!.H)$. We now show that these two conditions are always satisfied.

First of all, we have

$$H.i = H.i.j.i = H.t.H.i = t.MH.H.i = t.\Gamma.H.i = r.t.H.i = r.i.j.i = r.i,$$

showing that a hyperdeformation retraction is automatically a strong deformation retraction.

Secondly, to show $Mj.H = \alpha.(j, M!.H)$, we calculate:

$$\begin{aligned} Mi.Mj.H &= M(i.j).H \\ &= M(t.H).H \\ &= Mt.MH.H \\ &= Mt.\Gamma.H \end{aligned}$$

$$
\begin{aligned}
&= \alpha.(t, M!).H \\
&= \alpha.(t.H, M!.H) \\
&= \alpha.(i.j, M!.H) \\
&= Mi.\alpha.(j, M!.H).
\end{aligned}
$$

Since Mi is monic (even split monic), this proves the claim. We leave it to the reader to verify that this construction is functorial and induces an isomorphism of cofibred structures. □

Corollary 4.1 *HDRs admit a vertical composition, given by vertical composition of coalgebras for an AWFS. Explicitly, the composition of two HDRs $i_0 : A \to B$, $i_1, B \to C$ is given by $i_1.i_0$ with inverse map $j_0.j_1$ and deformation*

$$H_1 * H_0 := \mu.(H_1, Mi_1.H_0.j_1) \tag{4.6}$$

Proof We only need to verify that the given formula and inverse represents composition of coalgebras, as defined by the formula (2.21) above. Expanding this formula for $h := i_1.i_0$ yields:

$$
\begin{aligned}
&(H_1, j_1) * (H_0, j_0) \\
&= \mu_h.E(E(1_A, i_1).(H_0, j_0), 1_C).(H_1, j_1) \\
&= (\mu.(p_1, p_1.p_2), p_2.p_2).(M1_C.p_1, (Mi_1.p_1, p_2).(H_0, j_0).p_2).(H_1, j_1) \\
&= (\mu.(p_1, p_1.p_2), p_2.p_2).(H_1, (Mi_1.H_0, j_0).j_1) \\
&= (\mu.(H_1, Mi_1.H_0.j_1), j_0.j_1)
\end{aligned}
$$

whence the statement follows. □

Similarly, we have the rest of the structure of a double category, whose vertical morphisms are HDRs. In the rest of this section, we will study this structure in more depth and using HDRs rather than coalgebras. If $(i' : A' \to B', j', H')$ and $(i : A \to B, j, H)$ are HDRs, then a *morphism of HDRs* is defined as a pair of maps $f : A' \to A$ and $g : B' \to B$ such that

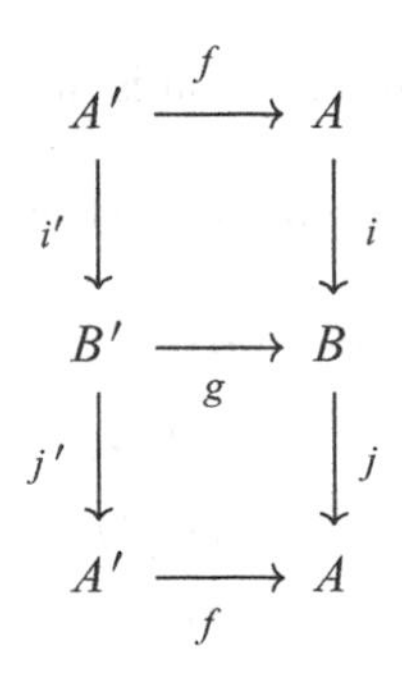

commutes and $Mg.H' = H.g : B' \to MB$.

Using the dual of Definition 2.14, we can summarise this section:

Corollary 4.2 *Hyperdeformation retracts define a discretely cofibred double category* $\mathbb{HDR}$ *over* $\mathcal{E}$ *which is isomorphic to the double category of coalgebras.*

Proof Using the dual of Propositions 2.7 and 4.3, it is enough to show that the correspondence of Proposition 4.3 is induced by a double functor L-$\mathbb{C}\mathbf{oalg} \to \mathbb{HDR}$ over $\mathbf{Sq}(\mathcal{E})$ which is full on squares. The functor exists by construction of the vertical composition, and the fact that a morphism of coalgebras is a morphism of HDRs. To see that it is full on squares, means to see that a morphism of HDRs is automatically a morphism of coalgebras, which also follows from the above definition. □

4.2.1 Hyperdeformation Retracts are Coalgebras

In this subsection and onwards, we denote by **HDR** the vertical part, as in Proposition 2.3, of the double category of HDRs, i.e. the category of HDRs and morphisms of HDRs. We first consider the *codomain* functor:

$$\text{cod} : \mathbf{HDR} \to \mathcal{E}. \tag{4.7}$$

The following facts will be helpful in Proposition 4.4 below.

Lemma 4.1 *If* (i, j, H) *is an HDR, then*

$$\begin{array}{ccc} A & \xrightarrow{i} & B \\ {\scriptstyle i}\downarrow & & \downarrow{\scriptstyle H} \\ B & \xrightarrow[r]{} & MB \end{array}$$

is a pullback. In particular, i *is the equalizer of the pair* $r, H : B \to MB$.

Proof If $i : A \to B$ is an HDR, as witnessed by $j : B \to A$ and $H : B \to MB$, then

$$\begin{array}{ccccc} A & \xrightarrow{i} & B & \xrightarrow{j} & A \\ {\scriptstyle i}\downarrow & & \downarrow{\scriptstyle H} & & \downarrow{\scriptstyle i} \\ B & \xrightarrow[r]{} & MB & \xrightarrow[t]{} & B \end{array}$$

exhibits i as a retract of H. Since H is monic ($s.H = 1$), Lemma 4.2 (see below) tells us that the left hand square is a pullback. □

Lemma 4.2 *If the commutative diagram*

$$\begin{array}{ccccc} A & \longrightarrow & C & \longrightarrow & A \\ \downarrow{\scriptstyle m} & & \downarrow{\scriptstyle n} & & \downarrow{\scriptstyle m} \\ B & \longrightarrow & D & \longrightarrow & B \end{array}$$

exhibits m as a retract of n and n is a monomorphism, then the left hand square is a pullback.

Proof This is dual to Lemma C.1 in the Appendix. A reference is given there. □

Corollary 4.3 *Suppose* (i', j', H'), (i, j, H) *are HDRs and* $g : B' \to B$ *satisfies* $Mg.H' = H.g$. *Then there is a unique* $f : A' \to A$ *such that* (f, g) *is a morphism of HDRs.*

Proof The unique f is defined by virtue of Lemma 4.1. It remains to verify that square $j' \to j$ between the retracts commutes:

$$\begin{aligned} j.g &= j.i.j.g = j.t.H.g = t.Mj.H.g \\ &= t.Mj.Mg.H' = t.M(j.g).H' \\ &= j.g.t.H' = j.g.i'.j' = j.i.f.j' = f.j' \end{aligned}$$

So the conclusion follows. □

The following proposition combines the previous two observations and will be used for developing the Frobenius construction (Lemma 5.2) and for developing the theory of HDRs in simplicial sets (e.g., Theorem 10.1).

Proposition 4.4 *The functor* (4.7) *is comonadic, where the corresponding comonad is just the Moore functor* $M : \mathcal{E} \to \mathcal{E}$ *with comonad structure* (s, Γ). *Specifically*

$$\mathrm{cod} : \mathbf{HDR} \to M\text{-}\mathbf{Coalg}, \tag{4.8}$$

which sends (i, j, H) *to* $(B, H : B \to MB)$, *is an equivalence of categories.*

Proof By Definition 4.2, the functor (4.8) has the correct codomain, i.e. it maps into M-coalgebras. By Corollary 4.3, the functor is full and faithful. Further, Lemma 4.1 and its proof imply that it is essentially surjective. So we have an equivalence of categories. □

We can use the previous result to prove the following:

Corollary 4.4 *The category of HDRs has pullbacks.*

Proof By Proposition 4.4, the functor (4.7) creates limits preserved by M. So it follows because M preserves pullbacks. □

To end this section, we record the following fact about HDRs, which holds when the unit $r_X : X \to MX$ is a *cartesian* natural transformation, i.e. all naturality squares are pullbacks. Note this happens to be true for simplicial sets—but it is not an assumption of our theory.

Proposition 4.5 *When* $r_X : X \to MX$ *is a* cartesian *natural transformation, every morphism of HDRs is a pullback square.*

Proof If

$$\begin{array}{ccc} B' & \xrightarrow{f} & B \\ {\scriptstyle i'}\downarrow & & \downarrow{\scriptstyle i} \\ A' & \xrightarrow[g]{} & A \end{array}$$

is (the top part of) a morphism of HDRs, then it fits into a commutative cube:

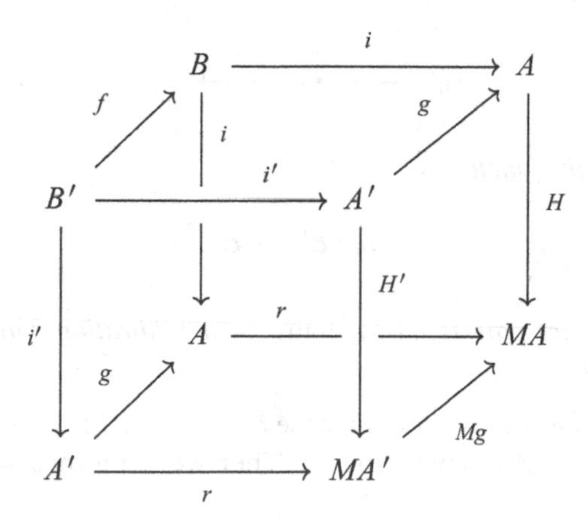

In this cube front and back faces are pullbacks (by Lemma 4.1), as is the bottom face (because r is a cartesian natural transformation). Therefore the top face is a pullback as well. □

4.2.2 Hyperdeformation Retracts are Bifibred

In the previous subsection, we have studied the codomain functor $\mathrm{cod} : \mathbf{HDR} \to \mathcal{E}$ and obtained the main result that it is comonadic. Next, we will turn to the *domain* functor $\mathrm{dom} : \mathbf{HDR} \to \mathcal{E}$.

Definition 4.3 A morphism of HDRs will be called a *cartesian morphism of HDRs* if also the bottom part, i.e. the square for j' and j, is a pullback.

Definition 4.4 A morphism of HDRs will be called a *cocartesian morphism of HDRs* if the square for i', i and the square for j', j are pushouts.

The main result of this subsection is the following:

Proposition 4.6 *The domain functor*

$$\mathrm{dom} : \mathbf{HDR} \to \mathcal{E}$$

is a bifibration *(see Chap. 2), whose cartesian morphisms are given by Definition 4.3, and whose cocartesian morphisms are given by Definition 4.4. Moreover, this bifibration satisfies the Beck-Chevalley condition (see Box 2.1 on p. 15).*

Before proving the proposition, we prove the following two lemmas:

Lemma 4.3 *Suppose* $\mathbf{r}$ *is the universal retract, in other words the category*

$$\bullet_0 \xrightarrow{i} \bullet_1 \xrightarrow{j} \bullet_0$$

with $j.i = 1$. *Then the functor*

$$\mathrm{ev}_0 : \mathcal{E}^{\mathbf{r}} \to \mathcal{E}$$

which sends a retract pair (i, j) *to* $\mathrm{dom}\, i$ *is a bifibration satisfying the Beck-Chevalley condition.*

Proof To prove that it is a fibration, suppose $i : A \to B$, $j : B \to A$ is a retract pair and suppose $f : A' \to A$ is any morphism. Then we can form a double pullback

$$\begin{array}{ccc} A' & \xrightarrow{f} & A \\ {\scriptstyle i':=(1,i.f)}\downarrow & & \downarrow{\scriptstyle i} \\ B' & \xrightarrow{g} & B \\ {\scriptstyle j'}\downarrow & & \downarrow{\scriptstyle j} \\ A' & \xrightarrow{f} & A \end{array}$$

resulting in a morphism of retract pairs $(i', j') \to (i, j)$. It is enough to verify that this morphism is cartesian over f, which is very easy.

Similarly, for a pair $(i' : A' \to A, j' : B' \to B)$ and a morphism $f : A' \to A$, the double pushout diagram

$$\begin{array}{ccc} A' & \xrightarrow{f} & A \\ {\scriptstyle i'}\downarrow & & \downarrow{\scriptstyle i} \\ B' & \xrightarrow{g} & B \\ {\scriptstyle j'}\downarrow & & \downarrow{\scriptstyle [f.j',1]} \\ A' & \xrightarrow{f} & A \end{array}$$

yields a retract pair $(i, [f.j', 1])$ and it is easy to see that this morphism of retract pairs satisfies the universal property of a cocartesian morphism over f.

For the Beck-Chevalley condition we prove the '$\Rightarrow$' direction of the definition explained in Box 2.1 on p. 15. So consider the situation of (2.5) for the case at hand, i.e. a diagram:

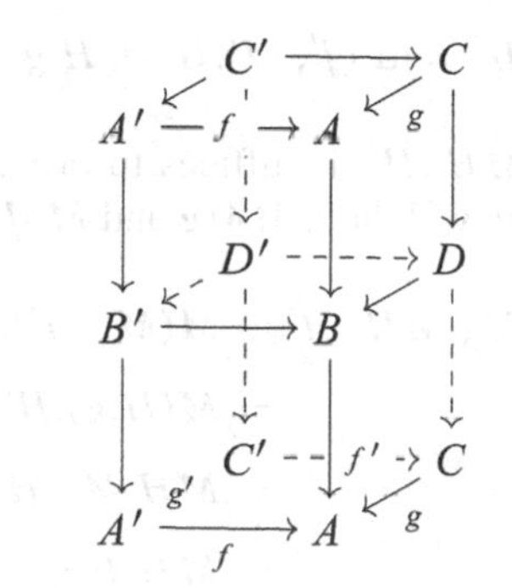

where the bottom and top squares pullbacks, the back (vertical) squares form a double pullback, and the right-hand vertical squares form a double pushout. It needs to be shown that if the front vertical squares are a double pullback, then the left vertical squares are a double pushout. But this follows immediately from the assumption on $\mathcal{E}$ stated at the beginning of Chap. 2. □

Lemma 4.4 *Suppose we have a morphism of retract pairs $(i', j') \to (i, j)$ given by $f : A' \to A$, $g : B' \to B$. Then if this morphism is*

(i) a cartesian morphism of retract pairs, and $H : B \to MB$ gives (i, j) the structure of an HDR, then there is a unique HDR structure on the pair (i', j') such that the cartesian morphism is a cartesian morphism of HDRs.

(ii) a cocartesian morphism of retract pairs, and $H : B' \to MB'$ gives (i', j') the structure of an HDR, then there is a unique HDR structure on the pair (i, j) such that the cocartesian morphism is a cocartesian morphism of HDRs.

Proof

(i) Since any HDR structure H' that would be a witness to the claim makes

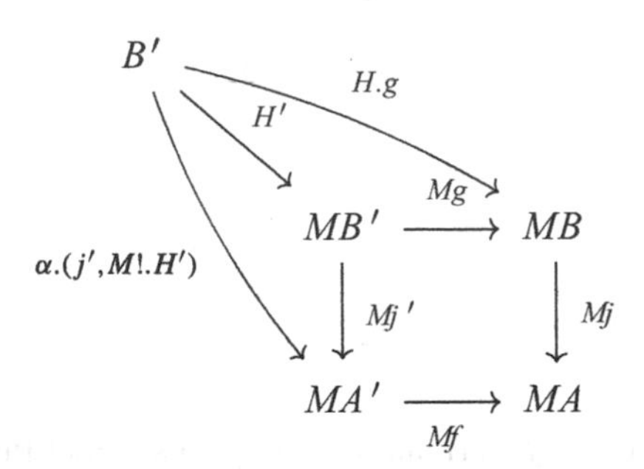

commute (see the proof of Proposition 4.3), it must be unique because M preserves pullback squares. It remains to see that H' can be defined in this way by setting

$$H' = (\alpha.(j', M!.H.g), H.g).$$

To check that $\Gamma.H' = MH'.H'$, it suffices to prove that both sides become equal upon postcomposing with both MMg and MMj'. But we have:

$$\begin{aligned} MMg.MH'.H' &= M(Mg.H').H' \\ &= M(H.g).H' \\ &= MH.Mg.H' \\ &= MH.H.g \\ &= \Gamma.H.g \\ &= \Gamma.Mg.H' \\ &= MMg.\Gamma.H' \end{aligned}$$

and

$$\begin{aligned} MMj.\Gamma.H' &= \Gamma.Mj.H' \\ &= \Gamma.\alpha(j', M!.H.g) \\ &= M\alpha.(\alpha.(j', M!.H.g), \Gamma.M!.H.g) \\ &= M\alpha.(Mj.H', MM!.\Gamma.H.g) \end{aligned}$$

$$
\begin{aligned}
&= M\alpha.(Mj.H', MM!.MH.H.g) \\
&= M\alpha.(Mj.H', MM!.MH.Mg.H') \\
&= M(\alpha(j, M!.H.g)).H' \\
&= M(Mg.H').H' \\
&= MMg.MH'.H'.
\end{aligned}
$$

Note we have used an identity from the proof of Proposition 4.3 here. Lastly, we verify:

$$
\begin{aligned}
t.H' &= t.(\alpha.(j', M!.H.g), H.g) \\
&= (t.\alpha.(j', M!.H.g), t.H.g) \\
&= (j', i.j.g) \\
&= (1, i.f).j'
\end{aligned}
$$

(ii) This is Proposition 4.2.

□

We can now prove the above stated proposition:

Proof of Proposition 4.6 By the previous two lemmas, it remains to show that when (i_0, j_0, H_0), (i_1, j_1, H_1) and (i_2, j_2, H_2) are HDRs such that we have a composite

$$
\begin{array}{ccccc}
A_2 & \xrightarrow{f'} & A_1 & \xrightarrow{f} & A_0 \\
{\scriptstyle i_2}\downarrow & & {\scriptstyle i_1}\downarrow & & {\scriptstyle i_0}\downarrow \\
B_2 & \xrightarrow{g'} & B_1 & \xrightarrow{g} & B_0 \\
{\scriptstyle j_2}\downarrow & & {\scriptstyle j_1}\downarrow & & {\scriptstyle j_0}\downarrow \\
A_2 & \xrightarrow{f'} & A_1 & \xrightarrow{f} & A_0
\end{array}
$$

which is a morphism of HDRs, then:

(i) If the right morphism of retract pairs is a cartesian morphism of HDRs, the left one is automatically a morphism of HDRs
(ii) If the left morphism is a cocartesian morphism of HDRs, then the right one is automatically a morphism of HDRs.

(i) follows again by taking projections:

$$
Mg.Mg'.H_2 = M(g.g').H_2 = H_0.(g.g') = (H_0.g).g' = Mg.(H_1.g')
$$

and

$$\begin{aligned} Mj_1.Mg'.H_2 &= M(f'.j_2).H_2 \\ &= M(f').\alpha.(j_2, M!.H_2) \\ &= \alpha.(f'.j_2, M!.H_2) \\ &= \alpha.(f'.j_2, M!.H_0.g.g') \\ &= \alpha.(j_1, M!.H_0.g).g' \\ &= \alpha.(j_1, M!.H_1).g' \\ &= Mj_1.H_1.g' \end{aligned}$$

(ii) follows again from Proposition 4.3, since the property is easy to verify for coalgebras using Proposition 2.5.

Observe that the Beck-Chevalley condition is now inherited from Lemma 4.3. □

The following is a first consequence of the more abstract approach we have taken so far:

Corollary 4.5 *In the category of HDRs, the pullback of a cartesian morphism of HDRs along a morphism of HDRs exists and is a cartesian morphism of HDRs.*

Proof This is a direct consequence of the fact that dom : **HDR** $\to \mathcal{E}$ is a Grothendieck fibration (and that $\mathcal{E}$ has pullbacks). □

4.3 Naive Fibrations

In this section, we will derive an alternative characterisation of the R-algebras associated to a category with Moore structure. We will show that R-algebras can be identified, as a notion of fibred structure, with maps that come equipped with a property that resembles a path-lifting property. This is analogous to the explicit characterisation of *cloven $\mathcal{R}$-maps* found by Van den Berg and Garner [19] for path object categories (see Proposition 6.1.5 loc. cit.).

Recall (from Sects. 4.1.1 and 4.1.3) that the monad for the AWFS defined in this section is given by $Rp = s.p_1$, so that algebras are fillers:

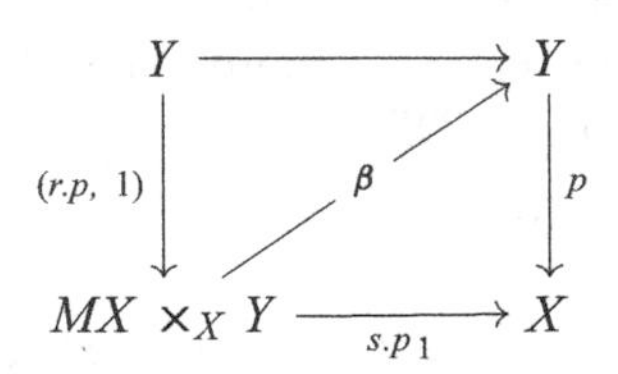

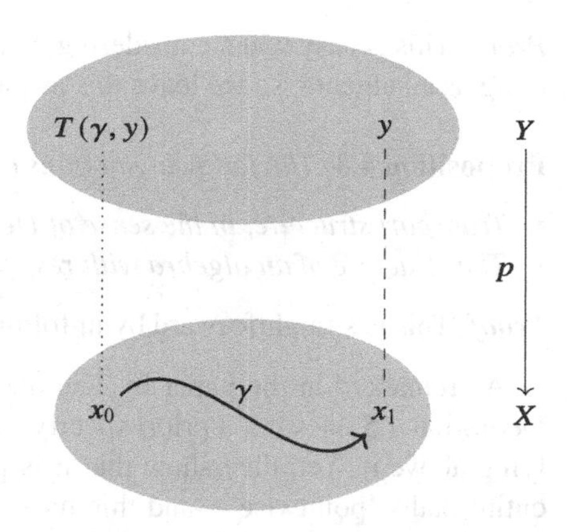

Fig. 4.2 Transport structure for p

satisfying an additional unit and associativity condition.

Since R-algebras have more structure than cloven $\mathcal{R}$-maps (namely, the unit and associativity condition), the alternative characterisation in terms of a path lifting property can be expected to meet more structural conditions. Yet the idea of the correspondence, as well as the further characterisation as *naive fibrations* originates from Van den Berg and Garner.

We have the following definition:

Definition 4.5 Let $p : Y \to X$ be any map. To equip p with *transport structure* means specifying a map

$$T : MX \times_X Y \to Y$$

where $MX \times_X Y$ is the pullback of t and p, with $p.T = s.p_1$, $T.(r.p, 1) = 1$ and such that

$$T.(\mu.(p_1.p_1, p_2.p_1), p_2) = T.(p_1.p_1, T.(p_2.p_1, p_2)) \, : \, (MX \times_X MX) \times_X Y \to Y.$$

where the first pullback is the pullback of t and s (the domain of μ) and the second of $t.p_1$ and p.

The above definition is illustrated in Fig. 4.2. Intuitively, an endpoint y in a space Y lying above a base space X is *transported* back along a path γ in the base space.

Proposition 4.7 *The function which associates to every* $p : Y \to X$ *the set*

$$\{T : MX \times_X Y \to Y \mid T \textit{ is a transport structure for } p\}$$

can be extended to a fibred structure

$$(\mathcal{E}^{\to}_{\text{cart}})^{op} \to \mathcal{S}ets$$

Proof This is easy when considering that a transport structure amounts to the same thing as an algebra, so we leave this as part of Proposition 4.8. □

Proposition 4.8 *The following notions of fibred structure are isomorphic:*

- *Transport structure, in the sense of Definition 4.5,*
- *The structure of an algebra with respect to $R = s.p_1$.*

Proof This is straightforward by unfolding the definition of an algebra. □

As remarked in the paper by Van den Berg and Garner, the lifting property of Definition 4.5 does not a priori specify that it is possible to lift a path to a full path lying above it. Yet, they show that it is possible to use the path contraction to lift entire paths 'pointwise'—and this turns out to work also in the setting presented here. We first give a definition of this path-lifting property for our setting, which again takes more structural axioms than in the former paper.

Definition 4.6 A map $p : Y \to X$ together with an arrow

$$L : MX \times_X Y \to MY$$

in $\mathcal{E}$ is said to be a *naive fibration*when it satisfies the conditions:

(i) $(Mp, t).L = 1$;
(ii) $L.(r.p, 1) = r$;
(iii) $L.(\mu.(p_1.p_1, p_2.p_1), p_2) = \mu.(L.(p_1.p_1, s.L.(p_2.p_1, p_2)), L.(p_2.p_1, p_2))$;
(iv) $\Gamma.L = ML.p_1.\delta_p$.

A naive fibration $p : Y \to X$ can be thought of as a map between spaces, such that whenever one has a path γ in the base X and a point y in Y which lies its end point $x_1 := t(\gamma)$, then one can construct a full path γ^* in MY lying above γ. This is illustrated in Fig. 4.3.

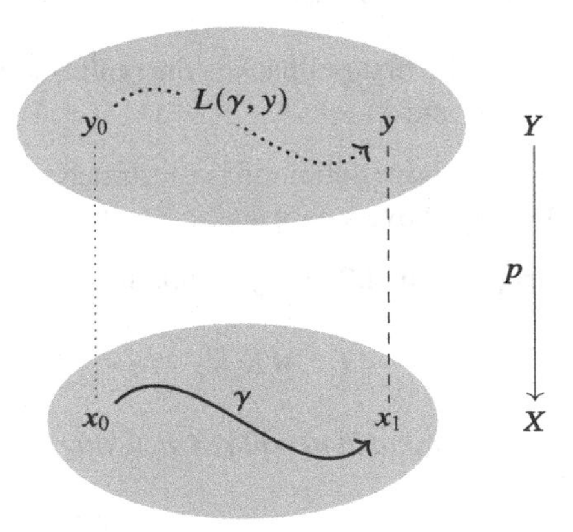

Fig. 4.3 A naive fibration p

Looking at Fig. 4.3, (i) ensures that $L(\gamma, y)$ lies above γ under p; (ii) that trivial paths are lifted to trivial paths; (iii) that lifting respects path composition; and (iv) that the contraction of γ specified by Γ lifts, pointwise under ML, to the contraction of $L(\gamma, y)$. The properties (iii) and (iv) differentiate naive fibrations from the path-lifting property established in the Proposition 6.1.5 of Van den Berg-Garner [19] mentioned at the beginning of this section.

Proposition 4.9 *The function which associates to every $p : Y \to X$ the set*

$$\{L : MX \times_X Y \to MY \mid (p, L) \text{ is a naive fibration}\}$$

can be extended to a fibred structure

$$\mathbf{nFib} : (\mathcal{E}^{\to}_{\text{cart}})^{\text{op}} \to \mathcal{S}ets$$

Proof Again, we leave this as part of their characterisation in Proposition 4.10. □

Proposition 4.10 *Let $p : Y \to X$ be a map. If L specifies a naive fibration structure on p, then $T = s.L$ is a transport structure on p. And if T is a transport structure on p, then $L = MT.p_2.\delta_p$ turns p into a naive fibration. These operations are mutually inverse and define an isomorphism between the following notions of fibred structure:*

- *Transport structure,*
- *Naive fibrations.*

Proof Suppose L satisfies the conditions (i)–(iv) of Definition 4.6, and let $T = s.L$. Then $p.T = p.s.L = s.Mp.L = s.p_1$, $T.(r.p, 1) = s.L.(r.p, 1) = s.r = 1$, and

$$\begin{aligned} T.(\mu.(p_1.p_1, p_2.p_1), p_2) =& s.L.(\mu.(p_1.p_1, p_2.p_1), p_2) \\ =& s.\mu.(L.(p_1.p_1, s.L.(p_2.p_1, p_2)), L.(p_2.p_1, p_2)) \\ =& s.L.(p_1.p_1, s.L.(p_2.p_1, p_2)) \\ =& T.(p_1.p_1, T.(p_2.p_1, p_2)), \end{aligned}$$

so T is a transport structure. In addition,

$$MT.p_1.\delta_p = Ms.ML.p_1.\delta_p = Ms.\Gamma.L = L,$$

so L can be reconstructed from T.

Conversely, suppose T is a transport structure on p, and define

$$L = MT.\nu.(\Gamma.p_1, \alpha.(p_2, M!.p_1))$$

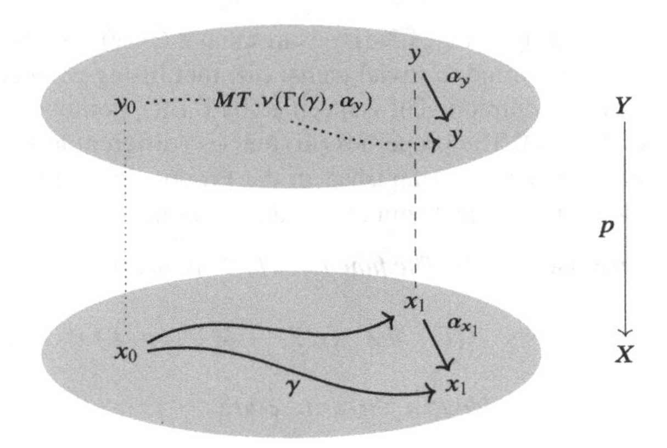

Fig. 4.4 Deriving lifting from (pointwise) transport. The paths α_y and α_{x_1} are constant paths of the same length as γ. The bottom triangle is the path $\Gamma(\gamma)$, contracting γ to the trivial path on x_1. Each 'copy' of y on α_y is transported along a final segment of γ lying transversal to $\Gamma(\gamma)$ (see Box 4.1)

where ν is given by the same mediating isomorphism as (4.4). This definition is illustrated in Fig. 4.4. Our first aim is to show (i)–(iv). First:

$$\begin{aligned} t.L &= t.MT.\nu.(\Gamma.p_1, \alpha.(p_2, M!.p_1)) \\ &= T.t.\nu.(\Gamma.p_1, \alpha.(p_2, M!.p_1)) \\ &= T.(t.\Gamma.p_1, t.\alpha.(p_2, M!.p_1)) \\ &= T.(r.t.p_1, p_2) \\ &= T.(r.p.p_2, p_2) \\ &= T.(r.p, 1).p_2 \\ &= p_2 \end{aligned}$$

and

$$\begin{aligned} Mp.L &= Mp.MT.\nu.(\Gamma.p_1, \alpha.(p_2, M!.p_1)) \\ &= M(p.T).\nu.(\Gamma.p_1, \alpha.(p_2, M!.p_1)) \\ &= M(s.p_1).\nu.(\Gamma.p_1, \alpha.(p_2, M!.p_1)) \\ &= Ms.Mp_1.\nu.(\Gamma.p_1, \alpha.(p_2, M!.p_1)) \\ &= Ms.p_1.(\Gamma.p_1, \alpha.(p_2, M!.p_1)) \\ &= Ms.\Gamma.p_1 \\ &= p_1, \end{aligned}$$

and hence $(Mp, t).L = 1$.

Furthermore,

$$\begin{aligned}
L.(r.p, 1) &= MT.\nu.(\Gamma.p_1, \alpha.(p_2, M!.p_1)).(r.p, 1)\\
&= MT.\nu.(\Gamma.r.p, \alpha.(1, M!.r.p))\\
&= MT.\nu.(r.r.p, \alpha.(1, r.!.p))\\
&= MT.\nu.(r.r.p, \alpha.(1, r.!))\\
&= MT.\nu.(r.r.p, r)\\
&= MT.r.(r.p, 1)\\
&= r.T.(r.p, 1)\\
&= r,
\end{aligned}$$

so also the second condition for a lift is satisfied.

The following calculation shows the third condition:

$$\begin{aligned}
L.(\mu.(p_1.p_1, p_2.p_1), p_2) &= MT.\nu.(\Gamma.p_1, \alpha.(p_2, M!.p_1)).(\mu.(p_1.p_1, p_2.p_1), p_2)\\
&= MT.\nu.(\Gamma.\mu.(p_1.p_1, p_2.p_1), \alpha.(p_2, M!.\mu.(p_1.p_1, p_2.p_1)))\\
&=^1 MT.\nu.(\mu.(M\mu.\nu.(\Gamma.p_1.p_1, \alpha.(p_2.p_1, M!.p_1.p_1)), \Gamma.p_2.p_1),\\
&\qquad \mu.(\alpha.(p_2, M!.p_1.p_1), \alpha.(p_2, M!.p_2.p_1)))\\
&=^2 MT.\mu.(\nu.(M\mu.\nu.(\Gamma.p_1.p_1, \alpha.(p_2.p_1, M!.p_1.p_1)), \alpha.(p_2, M!.p_1.p_1)),\\
&\qquad \nu.(\Gamma.p_2.p_1, \alpha.(p_2, M!.p_2.p_1)))\\
&=^3 \mu.(MT.\nu.(M\mu.\nu.(\Gamma.p_1.p_1, \alpha.(p_2.p_1, M!.p_1.p_1)), \alpha.(p_2, M!.p_1.p_1)),\\
&\qquad MT.\nu.(\Gamma.p_2.p_1, \alpha.(p_2, M!.p_2.p_1)))\\
&=^4 \mu.(MT.\nu.(\Gamma.p_1.p_1, MT.\nu.(\alpha.(p_2.p_1, M!.p_1.p_1), \alpha.(p_2, M!.p_1.p_1))),\\
&\qquad L.(p_2.p_1, p_2))\\
&=^5 \mu.(MT.\nu.(\Gamma.p_1.p_1, \alpha.(T.(p_2.p_1, p_2), M!.p_1.p_1)), L.(p_2.p_1, p_2))\\
&= \mu.(MT.\nu.(\Gamma.p_1, \alpha.(p_2, M!.p_1)).(p_1.p_1, T.(p_2.p_1, p_2)),\\
&\qquad L.(p_2.p_1, p_2))\\
&= \mu.(L.(p_1.p_1, T.(p_2.p_1, p_2)), L.(p_2.p_1, p_2))\\
&= \mu.(L.(p_1.p_1, s.L.(p_2.p_1, p_2)), L.(p_2.p_1, p_2))
\end{aligned}$$

where at $=^1$ we have used the distributive law, at $=^2$ we have rewritten the equation of the form (4.5), at $=^3$ we have used naturality of μ. At $=^4$, we have used the definition of L and further that

$$MT.\nu.(M\mu.\nu.(Mp_1.Mp_1, Mp_2.Mp_1), Mp_2) =$$
$$MT.\nu.(Mp_1.Mp_1, MT.\nu.(Mp_2.Mp_1, Mp_2))$$

which is the image under M of the requirement on T with respect to μ. The step $=^5$ uses naturality of α (for the square with T). Then it is a matter of rewriting, and in the last step we use the equation established at the end of this proof.

For the fourth condition, we again calculate:

$$\begin{aligned} ML.\nu.(\Gamma.p_1, \alpha.(p_2, M!.p_1)) &= MMT.\nu.M(\Gamma.p_1, \alpha.(p_2, M!.p_1)).\nu.(\Gamma.p_1, \alpha.(p_2, M!.p_1)) \\ &= MMT.\nu.(M\Gamma.Mp_1, \alpha.(Mp_2, MM!.Mp_1)).\nu.(\Gamma.p_1, \alpha.(p_2, M!.p_1)) \\ &=^1 MMT.\nu.(M\Gamma.\Gamma.p_1, \alpha.(\alpha.(p_2, M!.p_1), MM!.\Gamma.p_1)) \\ &= MMT.\nu.(\Gamma.\Gamma.p_1, \Gamma.\alpha.(p_2, M!.p_1)) \\ &= MMT.\Gamma.\nu.(\Gamma.p_1, \alpha.(p_2, M!.p_1)) \\ &= \Gamma.MT.\nu.(\Gamma.p_1, \alpha.(p_2, M!.p_1)) \\ &= \Gamma.L \end{aligned}$$

where $=^1$ uses the axioms of the strength α with respect to Γ, and the rest are naturality conditions. This shows that L yields the structure of a naive fibration.

Finally, we have

$$\begin{aligned} s.L &= s.MT.\nu.(\Gamma.p_1, \alpha.(p_2, M!.p_1)) \\ &= T.s.\nu.(\Gamma.p_1, \alpha.(p_2, M!.p_1)) \\ &= T.(s.\Gamma.p_1, s.\alpha.(p_2, M!.p_1)) \\ &= T.(p_1, p_2) \\ &= T, \end{aligned}$$

showing that the operations are mutually inverse. □

To conclude this section, it will be helpful to spell out the notion of morphism between naive fibrations explicitly, where the definition is fixed by the notion of morphism between the underlying algebras.

If $(p : E \to B, L)$, $(p : E' \to B', L')$ are naive fibrations, then a *morphism of naive fibrations* $(p, L) \to (p', L')$ is a commutative square

$$\begin{array}{ccc} E' & \xrightarrow{g} & E \\ {\scriptstyle p'}\downarrow & & \downarrow{\scriptstyle p} \\ B' & \xrightarrow[f]{} & B \end{array} \qquad (4.9)$$

such that

$$L.(g.p_1, Mf.p_2) = Mg.L' \tag{4.10}$$

The following corollary then summarises this section (cf. Corollary 4.2):

Corollary 4.6 *With the inherited vertical composition of algebras, the discretely fibred concrete double category (over $\mathcal{E}$) of naive fibrations* $\mathbb{N}$**Fib** *is isomorphic to the double category of algebras.*

Proof Dual and analogous to the proof of Corollary 4.2. □

As remarked in 2.7, we may also refer to the mere category of naive fibrations **NFib**, or the notion of fibred structure **nFib**.

Chapter 5
The Frobenius Construction

The goal of this chapter is to prove a *Frobenius property* (see Box 5.1 below) for the AWFS constructed in the previous chapter in the case that the Moore structure comes equipped with a certain additional symmetry.

Such a symmetry was assumed throughout by Van den Berg and Garner [19], who show an analogous Frobenius property for their cloven weak factorisation systems (see Definition 3.3.3. loc.cit.). Like our previous results above, the symmetry imposed here needs to satisfy additional structure to yield an analogous result for categories with Moore structure.

As in [19], the symmetry comes down to the fact that it is possible to reverse paths by means of a 'twist map': a natural transformation $\tau : M \Rightarrow M$ subject to axioms which we will now elaborate. First, the natural transformation reverses paths as announced:

(1) For every X, $\tau_X : MX \to MX$ is an internal, idempotent identity-on-objects functor between the category MX (given by the Moore structure) and the opposite category on MX. In particular $s.\tau = t$ and $t.\tau = s$.

Further, we require compatibility with α:

$$\tau.\alpha = \alpha(1, \tau) \tag{5.1}$$

So far, these requirements are the same as in [19]. Our additional axioms can be motivated as follows. The twist map can be used to create a new, reversed path contraction:

$$\Gamma^* = \tau_M.M(\tau).\Gamma.\tau \tag{5.2}$$

This is illustrated in Fig. 5.1. Using (5.1), one can work out that the constant sides of the triangle can be identified as shown. Similar, one can work out that the two diagrams can be composed pointwise (or vertically), as illustrated in Fig. 5.2.

B. van den Berg, E. Faber, *Effective Kan Fibrations in Simplicial Sets*,
Lecture Notes in Mathematics 2321, https://doi.org/10.1007/978-3-031-18900-5_5

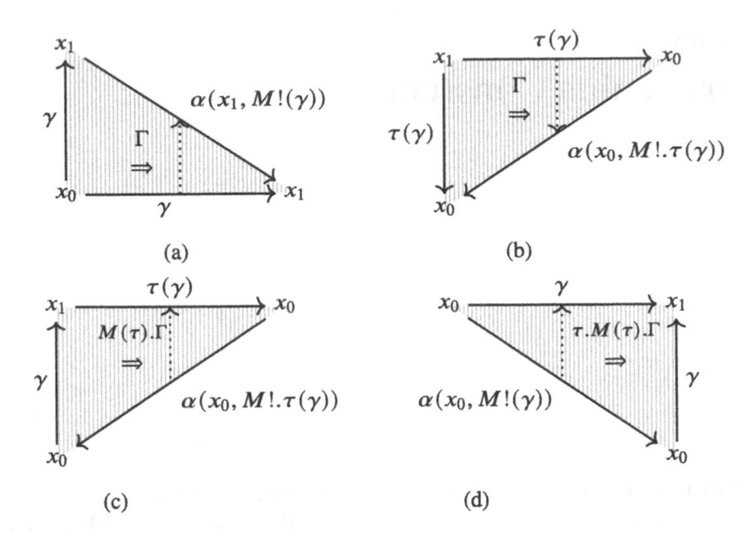

Fig. 5.1 Constructing the 'cocontraction' Γ^* using a twist map. **(a)** Γ. **(b)** $\Gamma.\tau(\gamma)$. **(c)** $M(\tau).\Gamma.\tau(\gamma)$. **(d)** $\tau.M(\tau).\Gamma.\tau(\gamma)$

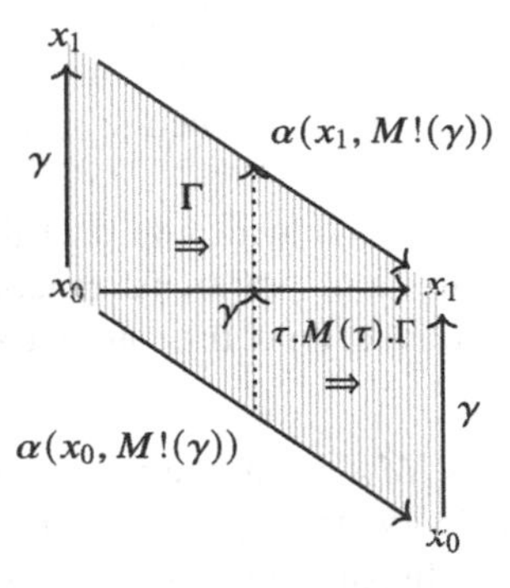

Fig. 5.2 Pointwise composition of the 'cocontraction' Γ^* with Γ

Of this pointwise composition, we will ask that it can be identified with the constant path from γ to itself of the same underlying length as itself:

$$M\mu.(\Gamma^*, \Gamma) = \alpha.(1, M!) : M \Rightarrow MM.$$

We will refer to this axiom as the *sandwich equation*, and it is a new requirement. Last, we need an axiom that is actually unrelated to the twist map, namely that composition is left (and right) cancellative: that is,

$$MX_t \times_s MX \xrightarrow[(1,\Delta)]{} MX_t \times_s (MX_{(t,s)} \times_{(t,s)} MX) \underset{\mu.(p_1, p_2.p_2)}{\overset{\mu.(p_1, p_1.p_2)}{\rightrightarrows}} MX$$

is an equalizer (and similarly for right cancellative). Since M preserves pullbacks, this remains an equalizer after applying M (so we can apply left cancellation pointwise). This axiom could have been part of the definition of Moore structure, but it so happened that we did not need it until this chapter.

Although the motivation of this chapter is symmetric Moore structures, the proofs presented are completely axiomatic with regard to Γ^* and Γ. Therefore we can merely assume a given natural transformation $\Gamma^* : M \Rightarrow MM$ with the following properties:

(1') $\Gamma^* : M \Rightarrow MM$ is a comonad satisfying the Moore structure axioms dual to to Γ (with s and t reversed);
(2) which satisfies the *sandwich equation*;
(3) and composition is left and right cancellative.

This slight abstraction will be called a *two-sided* Moore structure. The two definitions, of a two-sided Moore structure and a symmetric Moore structure, are summarised in Definitions A.2 and A.3.

5.1 Naive Left Fibrations

A two-sided Moore structure gives rise to a 'dual' AWFS on $\mathcal{E}$ induced by Γ^* instead of Γ. This yields a definition of *naive left fibration* (cf. Definition 4.6):

Definition 5.1 A map $p : Y \to X$ together with an arrow

$$L^* : MX \times_X Y \to MY,$$

where $MX \times_X Y$ refers to the pullback of p and s (instead of t), is said to be a *naive left fibration* when it satisfies the conditions:

(i) $(Mp, s).L^* = 1$;
(ii) $L^*.(r.p, 1) = r$;
(iii) $L^*.(\mu.(p_2.p_1, p_1.p_1), p_2) = \mu.(L^*.(p_1.p_1, t.L^*.(p_2.p_1, p_2)), L^*.(p_2.p_1, p_2))$;
(iv) $\Gamma^*.L^* = ML^*.(\Gamma^*.p_1, \alpha.(p_2, M!.p_1))$.

Remark 5.1 The terminology of left fibrations is adopted from the corresponding notions (due to Joyal) of left (and right) fibrations in the category of simplicial sets, see e.g. [35], chapter 2. Our notion of *effective* left and right fibration (developed in Chap. 6) for categories with Moore structure coincides with left and right fibrations in simplicial sets in this sense.

It follows from the previous that naive left fibrations are R-algebras for the AWFS induced by the dual comonad Γ^*. In particular, Corollary 4.6 gives:

Corollary 5.1 *Naive left fibrations have the structure of a discretely fibred double category* $\mathbb{N}$**LFib** *over* $\mathcal{E}$*, whose squares are given by commutative squares as in* (4.9).

Like previously, we may also refer to its structure as a category **NLFib** or notion of fibred structure **nLFib**.

For *symmetric* Moore structure, it turns out that the notion of left naive fibration is no different from the notion of naive fibration. This follows from the following proposition, which states that HDRs are the same for the two Moore structures given by Γ and Γ^*. The second part of the following proposition is stated so as to apply Proposition 2.7 immediately.

Proposition 5.1 *In a category with symmetric Moore structure, HDRs coincide for both Moore structures. That is, there is a double functor over* **Sq** $(\mathcal{E})$ *between the two (concrete) double categories of HDRs which is an isomorphism.*

Hence, there is an induced double functor between discretely fibred concrete double categories over $\mathcal{E}$*:*

$$\mathbb{N}\mathbf{Fib} \to \mathbb{N}\mathbf{LFib}.$$

This functor is full on squares, and induces an isomorphism between notions of fibred structure. Hence, this functor is an isomorphism.

Proof For the first part, we define a double functor as follows.

If $(i : A \to B, j, H^*)$ is a 'left-HDR', so for the dual structure given by Γ^*, then let $H := \tau_B.H^*$. We claim that (i, j, H) is an HDR. The first requirements are easy to check. For the condition on Γ, we have:

$$\begin{aligned}\Gamma.H = \Gamma.\tau.H^* &= M(\tau).\tau.\Gamma^*.H^* = M(\tau).\tau.M(H^*).H^* \\ &= M(\tau).M(H^*).\tau.H^* = M(\tau.H^*).\tau.H^* \\ &= M(H).H\end{aligned}$$

Functoriality with respect to squares is easy to see, and for vertical composition as given by (4.6) we have:

$$\begin{aligned}(\tau.H_1) * (\tau.H_0) &= \mu.(\tau.H_1, Mi_1.\tau.H_0.j_1) \\ &= \mu.(\tau.H_1, \tau.Mi_1.H_0.j_1) \\ &= \tau.\mu.(H_1, Mi_1.H_0.j_1) \\ &= \tau.(H_1 * H_0)\end{aligned}$$

Note that this proposition uses the first condition on τ entirely. Of course this argument dualizes, and clearly the two operations are inverse.

For the second part, it is a little exercise to show that the induced double functor between the categories of algebras satisfies the conditions, where algebra morphisms $MX \times_X \to Y$ for a naive fibration $p : Y \to X$ are sent to the precomposition with with

$$\tau \times_X Y : MX \times_X Y \to MX \times_X Y$$

(note, these are two different pullbacks, for s and t respectively or vice-versa). To do the exercise, one can apply the isomorphism to the HDR

$$(r.p, 1) : Y \to MX \times_X Y$$

with retraction given by the comultiplication. Then one can see what happens to the algebra morphism, which is defined as the diagonal filler with respect to this map. □

5.2 The Frobenius Construction

The equations (2)–(3) from the beginning of this chapter establish a relation between the two 'dual' AWFSs on $\mathcal{E}$ given by Γ and Γ^*. The following lemma exploits this in a way we will need for our formulation of the Frobenius construction in Proposition 5.2 below. In the presence of a two-sided Moore structure, we will now sometimes refer to naive fibrations as naive *right* fibrations, to emphasize which one we are talking about.

Lemma 5.1 *Suppose $p : Y \to X$ has the the structure of a naive right fibration $L : MX \times_X Y \to MY$, i.e. they satisfy the conditions of Definition 4.6. Then we also have:*

$$\Gamma^*.L = ML.(\Gamma^*.p_1, L).$$

for the dual Γ^*.

Dually, a naive left fibration $(p : Y \to X, L^*)$, satisfies

$$\Gamma.L^* = ML^*.(\Gamma.p_1, L^*).$$

Proof Note that by postcomposing (iv) in Definition 4.6 with Ms we obtain:

$$L = Ms.ML.(\Gamma.p_1, \alpha.(p_2, M!.p_1)).$$

By (pointwise) left cancellation and equation (iv) in Definition 4.6, it suffices to prove:

$$M\mu.(\Gamma^*.L, \Gamma.L) = M\mu.(ML.(\Gamma^*.p_1, L), ML.(\Gamma.p_1, \alpha.(p_2, P!.p_1)))$$

But we have

$$\begin{aligned} M\mu.(\Gamma^*.L, \Gamma.L) &= M\mu.(\Gamma^*, \Gamma).L \\ &= \alpha.(1, M!).L \\ &= \alpha.(L, M!.p_1), \end{aligned}$$

as well as

$$\begin{aligned} &M\mu.(ML.(\Gamma^*.p_1, L), ML.(\Gamma.p_1, \alpha.(p_2, M!.p_1))) = \\ &M\mu.(ML.(\Gamma^*.p_1, Ms.ML.(\Gamma.p_1, \alpha.(p_2, M!.p_1))), ML.(\Gamma.p_1, \alpha.(p_2, M!.p_1))) = \\ &ML.(M\mu.(\Gamma^*.p_1, \Gamma.p_1), \alpha.(p_2, M!.p_1)) = \\ &ML.(\alpha.(p_1, M!.p_1), \alpha.(p_2, M!.p_1)) = \\ &ML.\alpha.(1, M!.p_1) = \\ &\alpha.(L, M!.p_1). \end{aligned}$$

The second statement is dual to the first. □

The following proposition contains our definition of the *Frobenius construction* in the current context. As a property of an AWFS resulting from a Moore structure, the proposition is comparable to the *Frobenius property* of Van den Berg-Garner [19]. See also Box 5.1.

Box 5.1 The Frobenius Property

In recent literature, the Frobenius property is often formulated as 'the pullback of an $\mathcal{L}$-map along an $\mathcal{R}$-map is an $\mathcal{L}$-map' [4][8]. Here $\mathcal{L}$, $\mathcal{R}$ are the left and right classes of maps for an algebraic or non-algebraic weak factorisation system:

$$\begin{array}{ccc} D & \dashrightarrow & A \\ {\scriptstyle \in\mathcal{L}}\downarrow & & \downarrow{\scriptstyle i\in\mathcal{L}} \\ E & \xrightarrow[p\in\mathcal{R}]{} & B \end{array}$$

The property is of interest because it is closely related to *right-properness* of a model structure. This is the property that the factorisation system of

(continued)

Box 5.1 (continued)
trivial cofibrations and fibrations, as part of the model structure, satisfies the Frobenius property (see [8]). Further, it relates to dependent products (pushforward) for a model of type theory (as in Chap. 7 below). The Frobenius property for two-sided Moore structures is more abstract, since the class of right maps $\mathcal{R}$ in the above diagram is replaced by the right maps from a different, but closely related, AWFS. The actual Frobenius property is extracted as a consequence for *symmetric* Moore structures in Corollary 5.2.

Proposition 5.2 (Frobenius Construction) *Suppose (i, j, H) is an HDR and*

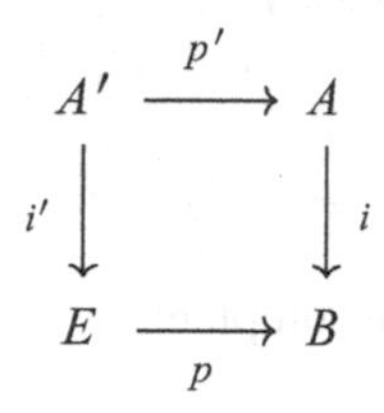

is a pullback square in which $p : E \to B$ is a naive left fibration. Then i' can be extended to an HDR such that the square becomes a morphism of HDRs.

Remark 5.2 The diagram above will not be a cartesian morphism of HDRs, in general.

Proof Let L^* be the naive left fibration structure on p. We have a map $j : B \to A$ and a homotopy $H : B \to MB$ with $j.i = 1, s.H = 1, t.H = i.j$ and $\Gamma.H = MH.H$. In addition, we have a pullback diagram of the form

$$\begin{array}{ccc} A' & \xrightarrow{p'} & A \\ {\scriptstyle i'}\downarrow & & \downarrow{\scriptstyle i} \\ E & \xrightarrow[p]{} & B. \end{array}$$

Write $H' = L^*.(H.p, 1) : E \to ME$. Then $s.H' = s.L^*.(H.p, 1) = p_2.(H.p, 1) = 1$ and

$$p.t.H' = p.t.L^*.(H.p, 1) = t.Mp.L^*.(H.p, 1) = t.p_1.(H.p, 1) = t.H.p = i.j.p,$$

and therefore there is a map $j' : E \to A'$ with $p'.j' = j.p$ and $i'.j' = t.H'$. We will first show that $j'.i' = 1$ and $MH'.H' = \Gamma.H'$.

To see $j'.i' = 1$, we calculate

$$\begin{aligned} i'.j'.i' &= t.H'.i' \\ &= t.L^*.(H.p, 1).i' \\ &= t.L^*.(H.p.i', i') \\ &= t.L^*.(H.i.p', i') \\ &= t.L^*.(r.i.p', i') \\ &= t.L^*.(r.p.i', i') \\ &= t.L^*.(r.p, 1).i' \\ &= t.r.i' \\ &= i' \\ &= i'.1. \end{aligned}$$

To prove $MH'.H' = \Gamma.H'$, we compute:

$$\begin{aligned} MH'.H' &= M(L^*.(H.p, 1)).L^*.(H.p, 1) \\ &= ML^*.M(H.p, 1).L^*.(H.p, 1) \\ &= ML^*.(MH.Mp, 1).L^*.(H.p, 1) \\ &= ML^*.(MH.Mp.L^*, L^*).(H.p, 1) \\ &= ML^*.(MH.p_1, L^*).(H.p, 1) \\ &= ML^*.(MH.H.p, L^*(H.p, 1)) \\ &= ML^*.(\Gamma.H.p, L^*.(H.p, 1)) \\ &= ML^*.(\Gamma.p_1, L^*).(H.p, 1) \\ &= \Gamma.L^*.(H.p, 1) \\ &= \Gamma.H'. \end{aligned}$$

Here we have used the identity of Lemma 5.1.

It remains to check that the square is a morphism of HDRs. However, we have $p'.j' = j.p$, by construction, and

$$\begin{aligned} Mp.H' &= Mp.L^*.(H.p, 1) \\ &= p_1.(H.p, 1) \\ &= H.p. \end{aligned}$$

□

Definition 5.2 A *Frobenius morphism* of HDRs is a morphism of HDRs as constructed in the proof of Proposition 5.2. So the domain is completely determined by an underlying HDR $(i : A \to B, j, H)$ (the codomain) and naive left fibration $p : E \to B$. It will be seen in Lemma 5.3 that Frobenius morphisms of HDRs are stable under pullback along morphisms of HDRs. This fact makes the proof of the main theorem of this chapter, Theorem 7.1, much more manageable.

Lemma 5.2 *The Frobenius construction of Proposition 5.2 is stable under composition of naive left fibrations p as well as composition of HDRs i.*

Proof Consider a picture as follows:

$$\begin{array}{ccccc} A_2 & \xrightarrow{q_1} & A_1 & \xrightarrow{q_0} & A_0 \\ {\scriptstyle i_2}\downarrow & & {\scriptstyle i_1}\downarrow & & \downarrow{\scriptstyle i_0} \\ E_2 & \xrightarrow[p_1]{} & E_1 & \xrightarrow[p_0]{} & E_0. \end{array}$$

Then we have

$$\begin{aligned} H_1 &= L^*_{p_0}.(H_0.p_0, 1) \\ H_2 &= L^*_{p_1}.(H_1.p_1, 1) \\ H_2^* &= L^*_{p_0.p_1}.(H_0.p_0.p_1, 1) \end{aligned}$$

In view of Proposition 4.4, it suffices to show $H_2^* = H_2$, which we can do as follows:

$$\begin{aligned} H_2^* &= L^*_{p_0.p_1}.(H_0.p_0.p_1, 1) \\ &= L^*_{p_1}.(L^*_{p_0}.(p_1, H_0.p_0.p_1), 1) \\ &= L^*_{p_1}.(L^*_{p_0}.(1, H_0.p_0).p_1), 1) \\ &= L^*_{p_1}.(H_1.p_1, 1) \\ &= H_2. \end{aligned}$$

Now consider a picture as follows:

$$\begin{array}{ccc} E_2 & \xrightarrow{p_2} & A_2 \\ {\scriptstyle i_1'}\downarrow & & \downarrow{\scriptstyle i_1} \\ E_1 & \xrightarrow{p_1} & A_1 \\ {\scriptstyle i_0'}\downarrow & & \downarrow{\scriptstyle i_0} \\ E_0 & \xrightarrow[p_0]{} & A_0 \end{array}$$

Then we have:

$$\begin{aligned} H_0' &= L_{p_0}^*.(H_0.p_0, 1) \\ H_1' &= L_{p_1}^*.(H_0.p_1, 1) \\ H_2' &= H_0' * H_1' \\ H_2^* &= L_{p_0}^*.((H_0 * H_1).p_0, 1) \end{aligned}$$

and we have to compare H_2' and H_2^*. So here we go:

$$\begin{aligned} H_2^* &= L_{p_0}^*.((H_0 * H_1).p_0, 1) \\ &= L_{p_0}^*.(\mu.(H_0.p_0, Mi_0.H_1.j_0.p_0), 1) \\ &= \mu.(L_{p_0}^*.(H_0.p_0, 1), L_{p_0}^*.(Mi_0.H_1.j_0.p_0, t.L_{p_0}^*(H_0.p_0, 1))) \\ &= \mu.(H_0', L_{p_0}^*.(Mi_0.H_1.j_0.p_0, t.H_0',)) \\ &= \mu.(H_0', L_{p_0}^*.(Mi_0.H_1.p_1.j_0', i_0'.j_0')) \\ &= \mu.(H_0', Mi_0'.L_{p_1}^*(H_1.p_1, 1).j_0') \\ &= \mu.(H_0', Mi_0'.H_1'.j_0') \\ &= H_0' * H_1' \\ &= H_2' \end{aligned}$$

and the proof is finished. □

Recall from Definition 5.2 that we call a morphism of HDRs arising from the Frobenius construction a *Frobenius morphism* of HDRs. The following lemma shows that Frobenius morphisms are stable under pullbacks. As mentioned above, the proof of Theorem 7.1 (specifically Lemma 7.1) is greatly simplified by this result. The property stated below is analogous to the property (of a weak factorisation system) of being *functorially Frobenius* in Van den Berg-Garner [19]. The lemma thus shows that this property also holds in the setting of an algebraic weak factorisation system for a category with (two-sided) Moore structure.

Lemma 5.3 (Pullback Stability of Frobenius Construction) *The Frobenius construction of Proposition 5.2 defines a* functor*:*

$$(-)^*(-) : \mathbf{NLFib} \times_{\mathcal{E}} \mathbf{HDR} \to \mathbf{HDR}$$

where the domain is the pullback of the domain functors to $\mathcal{E}$*.*

As a consequence, the pullback (Corollary 4.4) in **HDR** *of a Frobenius morphism along a morphism of HDRs is again a Frobenius morphism of HDRs.*

Proof Suppose $(a, b) : i_1 \to i_0$ is a morphism of HDRs, $(q_0, p_0) : i'_0 \to i_0$ is a Frobenius morphism of HDRs, and $(f, b) : p_1 \to p_0$ is a morphism of naive left fibrations, as in the solid part of the following diagram:

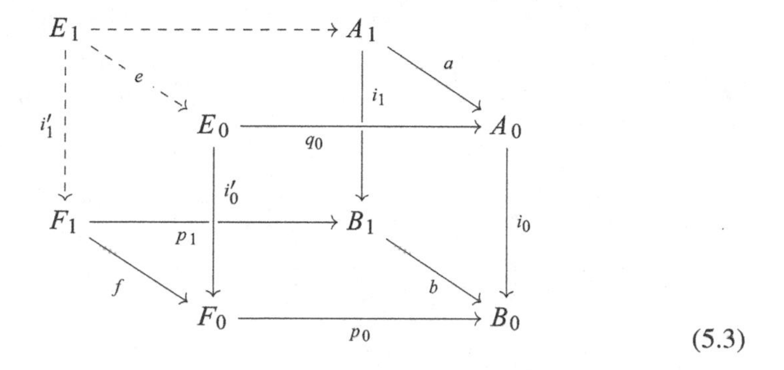

(5.3)

It is enough to prove that the Frobenius construction applied to the back square induces a unique morphism of HDRs $i'_1 \to i'_0$ on the left side of the cube.

So all that needs to be verified is that $(e, f) : i'_1 \to i'_0$ induced by the pullback is a morphism of HDRs. Denoting their respective HDR structure by H'_1, H'_0, and denoting the naive left fibrations by (p_0, L_{p_0}) and (p_1, L_{p_1}), we compute:

$$\begin{aligned} H'_0.f &= L^*_{p_0}.(H_0.p_0, 1).f = L^*_{p_0}.(H_0.p_0.f, f) \\ &= L^*_{p_0}.(H_0.b.p_1, f) = L^*_{p_0}.(Mb.H_1.p_1, f) \\ &= Mf.L^*_{p_1}.(H_1.p_1, 1) = Mf.H'_1 \end{aligned}$$

Where we have used that the bottom face is a morphism of naive left fibrations. By Corollary 4.3, it follows that the left face is a morphism of HDRs.

For the last statement, it is easy to see that the cube is a pullback square of morphisms of HDRs. □

As a corollary, we can rephrase the Frobenius construction as follows for *symmetric* Moore structures:

Corollary 5.2 (Frobenius for Symmetric Moore Structures) *In categories with symmetric Moore structure, there is a pullback functor*

$$(-)^*(-) : \mathbf{NFib} \times_{\mathcal{E}} \mathbf{HDR} \to \mathbf{HDR}$$

given by factoring the Frobenius construction of Lemma 5.3 through the isomorphism between naive left and right of Proposition 5.1.

Chapter 6
Mould Squares and Effective Fibrations

In this chapter, we connect the two algebraic weak factorisation systems coming from a dominance and a Moore structure to define the notion of *effective fibration*. We assume that $\mathcal{E}$ is a finitely cocomplete, locally cartesian closed category with Moore structure (recall the remarks in the introduction of Chap. 2). We assume further that $\mathcal{E}$ comes equipped with a dominance as in Chap. 3, such that:

- Σ is closed under binary unions of subobjects;
- Σ contains every initial arrow $0 \to A$ in $\mathcal{E}$.

As in Chap. 3, the coalgebras coming from the dominance are *effective cofibrations* and referred to as such. Lastly, we also make the combining assumption that every trivial Moore path

$$r_X : X \to MX$$

is an effective cofibration. This assumption can be thought of as the condition that the proposition expressing that a Moore path is trivial is cofibrant, i.e., contained in the dominance Σ. When r is a *cartesian* natural transformation, the condition is equivalent to saying that $r_1 : 1 \to M1$ is an effective cofibration. That is, it is enough that the proposition expressing that a Moore path has trivial *length* is cofibrant. In the case of simplicial sets, these last two are the case.[1]

Given a Moore structure and a dominance subject to these requirements, we have the following definition.

[1] We will define effective cofibrations in simplicial sets as locally decidable monomorphisms. It will follow easily that a Moore path having trivial length is locally decidable.

B. van den Berg, E. Faber, *Effective Kan Fibrations in Simplicial Sets*,
Lecture Notes in Mathematics 2321, https://doi.org/10.1007/978-3-031-18900-5_6

Definition 6.1 A *mould square* is a cartesian morphism of HDRs in the fibre above an effective cofibration, as in the following diagram:

$$\begin{array}{ccc} A' & \xrightarrow{i'} & B' \\ {\scriptstyle m}\downarrow & & \downarrow{\scriptstyle m'} \\ A & \xrightarrow{i} & B \end{array} \tag{6.1}$$

where i, i' are HDRs and m is an effective cofibration (so m' is also an effective cofibration).

A typical illustrative example of a mould square is given in Fig. 6.1. This example also serves to explain the choice of terminology. In the category of simplicial sets, there is a special type of mould square that we call *horn squares*. These are discussed in Chap. 12.

The following Lemma will be used in Definition 6.4 to define the *triple category* (see below) against which the notion of effective fibration will be defined.

Lemma 6.1 *The pullback of a mould square along an arbitrary morphism of HDRs is again a mould square.*

Proof This follows directly from Corollary 4.5 and the fact that effective cofibrations are stable under pullback. □

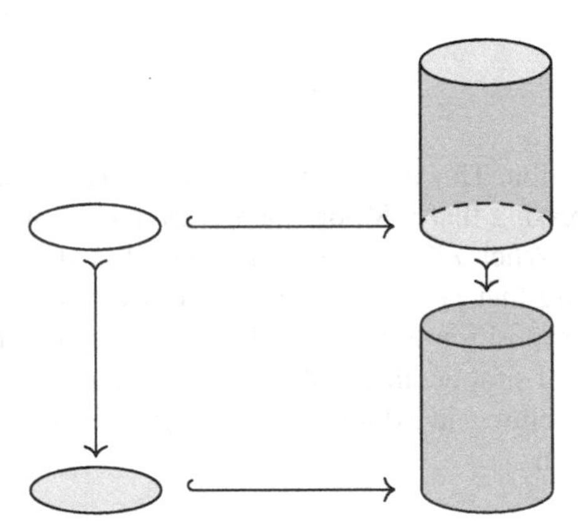

Fig. 6.1 An illustrative example of a mould square. The horizontal arrows indicate hyperdeformation retracts. On top the inclusion of a circle into the bottom end of a tube. On the bottom the inclusion of a disk into the bottom of a solid cylinder. The vertical arrows are cofibrations. The tube can be thought of as the upstanding walls of a type of 'mould' whose shape is the circle and whose base is the disk. The solid cylinder is the filled mould

Box 6.1 Triple Categories

Triple categories are not much harder to understand than double categories. We can define a *small* triple category as an internal category in the category of small double categories. This definition can be unfolded (and hence extended to include large triple categories) in the same way as in Sect. 2.2. Briefly, they extend double categories in that there is an additional type of 1-dimensional morphisms, *perpendicular* morphisms, and hence three different types of squares, between each pair of distinct 1-dimensional morphism types. In addition, *cubes* are morphisms between xy-squares which compose in the perpendicular direction, and come with additional 'pointwise' compositions for the horizontal and vertical direction. So a cube looks like:

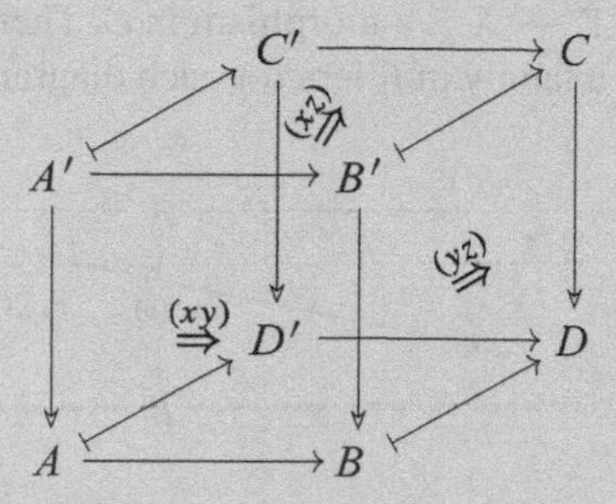

Note we are using the convention here that the three axes in 3-dimensional space are named as follows:

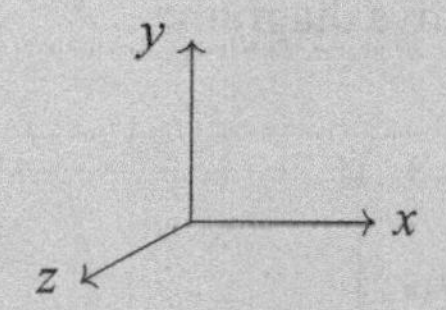

A cube admits composition from three different sides. The axioms guarantee that composition of a given combination of squares and cubes is independent of the order in which compositions are taken along the different dimensions. The standard example of a triple category is the category **Cube** $(\mathcal{E})$ for a category $\mathcal{E}$, where objects are objects in $\mathcal{E}$, all morphisms are given by morphisms in $\mathcal{E}$, all squares are commutative squares, and cubes are commutative cubes.

6.1 A New Notion of Fibred Structure

This section prepares the ground for Definition 6.4 in the next section. We introduce a definition of a family of 'fillers' for lifting problems which deviates from the right lifting structures defined in 2.2 for double categories. Namely, we introduce the notion of a right lifting structure with respect to a *triple category*. For brevity, we have put an introduction to triple categories in Box 6.1. In the end, it is only the definition of two specific triple categories (involving mould squares and effective cofibrations) that we will need to define effective (trivial) fibrations. The following definition spells out what it means to have a right lifting structure with respect to a triple category over **Cube** ($\mathcal{E}$).

Definition 6.2 Suppose $\mathbb{L}$ is a triple category and $I : \mathbb{L} \to$ **Cube** ($\mathcal{E}$) is a triple functor, and suppose $p : Y \to X$ is a morphism in $\mathcal{E}$. Then a *right lifting structure* for p with respect to I is a family of fillers for each diagram:

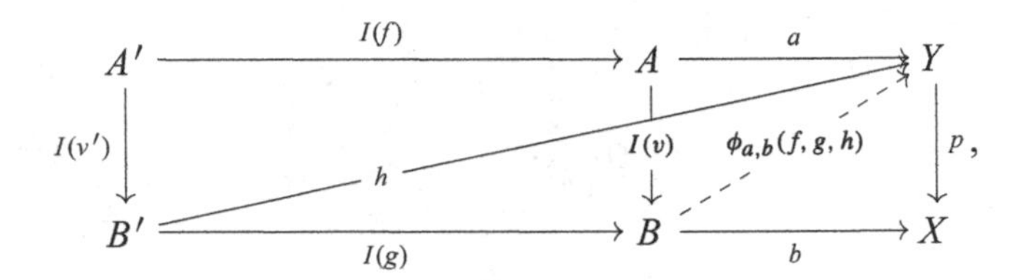

where the left-hand square is the image of an xy-square (see Box 6.1) under I, such that the following compatibility conditions hold:

Horizontal When (f', g'), (f, g) is a horizontally composable pair of xy-squares, and we are given a commutative diagram:

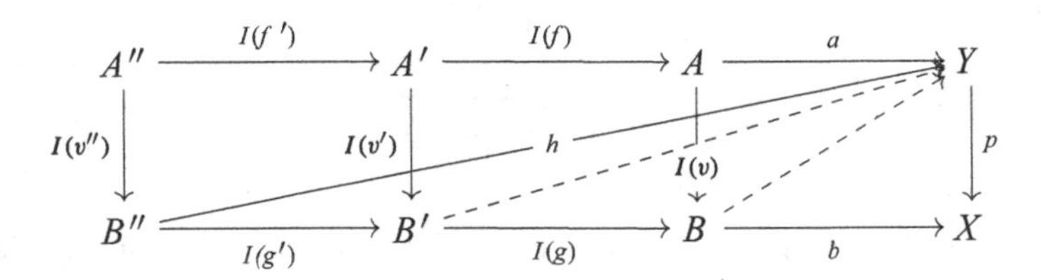

then we have:

$$\phi_{a,b}(f.f', g.g', h) = \phi_{a,b}(f, g, \phi_{a.I(f),b.I(g)}(f', g', h))$$

Vertical If we have a vertically composable pair of xy-squares and a diagram:

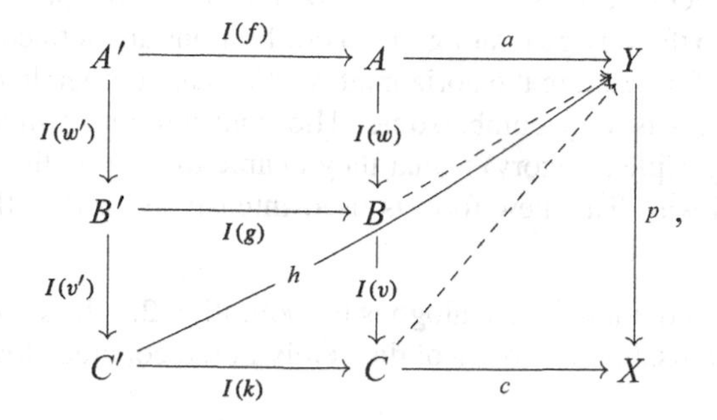

then we have:

$$\phi_{a,c}(f,k,h) = \phi_{\phi_{a,c.I(v)}(f,g,h.I(v')),c}(g,k,h)$$

Perpendicular For *cubes*, the condition asks that for the image of a cube between xy-squares:

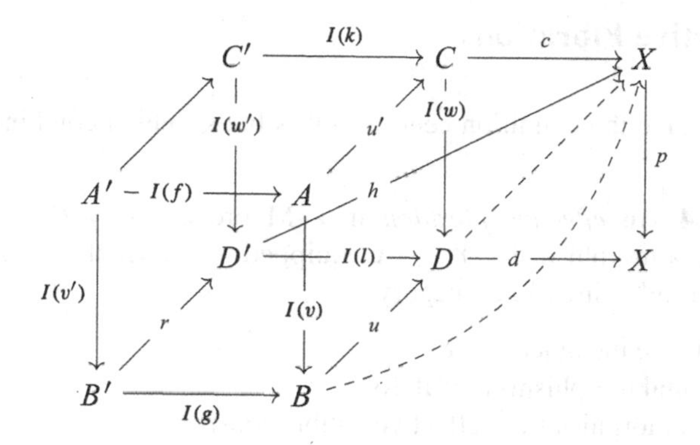

we have

$$u.\phi_{c,d}(k,l,h) = \phi_{d.u,c.u'}(f,g,h.r).$$

Remark 6.1 As for double categories, there is a symmetry in the definition of a triple category, namely the choice of 'top level' domain and codomain between cubes (the same goes for squares), which could be any of the three xy,yz,xz directions. In Definition 6.2 it is assumed that cubes are morphisms between xy-squares, just like the definition in Box 6.1. The definition of lifting structure takes this choice as a starting point. We do note that the definition of right lifting structure is symmetric in x and y, i.e., we could swap the horizontal and vertical morphisms.

Remark 6.2 The fillers in Definition 6.2 can also be given as ordinary diagonal fillers with respect to arrows $B' +_{A'} A \to B$ induced by the pushout of the square. This is similar to the way generating trivial cofibrations are defined by Gambino and Sattler [8]. Yet, formulating the horizontal, vertical and perpendicular conditions in terms of pushouts is very cumbersome. The contribution of mould squares (and the surrounding triple category) is that they enable to express these conditions in a straightforward way. This new form is used intensively in Part II of this book on simplicial sets.

The following definition is analogous to Definition 2.6. It is left to the reader to spell out the details. For the notion of discretely fibred concrete double category, see Definition 2.14.

Definition 6.3 There is a discretely fibred concrete double category over $\mathcal{E}$ of right lifting structures with respect to a triple category. In this category, squares are commutative squares satisfying the analogous compatibility condition (iii) of Definition 2.6 with respect to the triple category.

6.2 Effective Fibrations

We can now give the definition central to this book, which combines all previous chapters.

Definition 6.4 An *effective fibration* in a Moore category $\mathcal{E}$ equipped with a dominance is a morphism $p : Y \to X$ equipped with a right lifting structure with respect to the following triple category:

(i) Objects are the objects of $\mathcal{E}$.
(ii) Horizontal morphisms are HDRs.
(iii) Vertical morphisms are effective cofibrations.
(iv) Perpendicular morphisms are morphisms in $\mathcal{E}$.
(v) xy-squares are mould squares.
(vi) xz-squares are morphisms of HDRs.
(vii) yz-squares are morphisms of effective cofibrations, i.e. pullback squares.
(viii) Cubes are pullback 'squares' of a mould square along a morphism of HDRs (which always yields a mould square as per Lemma 6.1).

Note that cubes in this triple category are unique for a given boundary of a cube, which consists of six faces.

We denote the discretely fibred concrete double category (over $\mathcal{E}$) of effective fibrations by $\mathbb{E}$**ffFib** (see Definition 6.3). We may also refer to it as a category (**EffFib**) or notion of fibred structure (**effFib**) as explained in Sect. 2.6.

An illustration of an effective fibration, referring to the illustration of a mould square on p. 88, is given in Fig. 6.2. For a concrete example of the compatibility

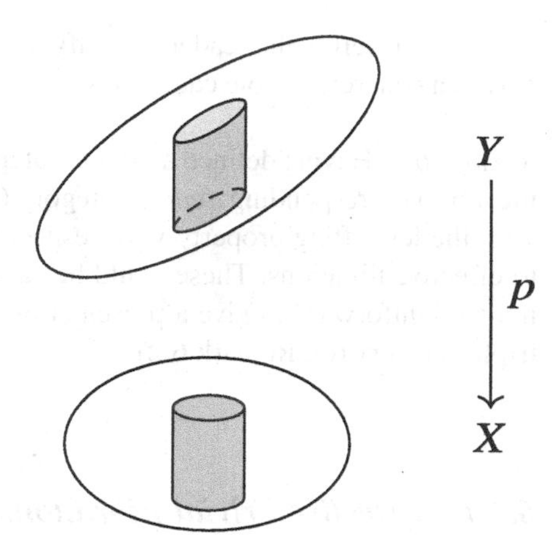

Fig. 6.2 Illustrative example of an (effective) fibration. The mould square (Fig. 6.1) defines a tube with bottom in Y (a 'mould'), lying above a solid cylinder in X. The fibration structure prescribes a way to fill the mould, that is compatible with other fillers (defined using a triple category)

conditions, it is best to look at simplicial sets directly, for which see Chaps. 11 and 12. For effective (Kan) fibrations in simplicial sets, the compatibility conditions category ultimately come down to a certain compatibility with respect to degeneracy maps (Theorem 12.2).

The following lemma can be used to motivate the terminology we have used for effective fibrations in relation to naive fibrations—we show that every effective fibration is a naive fibration. For the notion of a discretely fibred concrete double category, recall Definition 2.14. The lemma implies that there is an induced natural transformation between notions of fibred structure.

Lemma 6.2 *There is a double functor* $\mathbb{E}\mathbf{ffFib} \to \mathbb{N}\mathbf{Fib}$ *between the discretely fibred concrete double categories (over* $\mathcal{E}$*) of effective fibrations and naive fibrations.*

Proof We use the assumption from the beginning of this chapter, that every object is cofibrant, i.e. every $0 \to A$ is contained in the dominance. Suppose $p : Y \to X$ is an effective fibration. We will show that p can be equipped with a right lifting structure with respect to the double category of HDRs. Given an HDR $i : A \to B$, we can define the structure as the family of fillers:

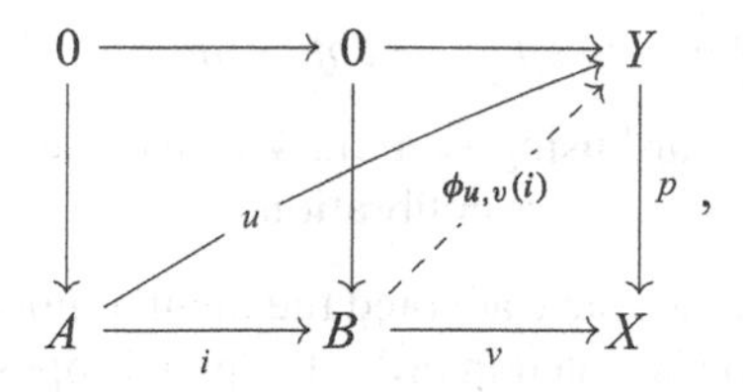

By our assumption, the square on the left is indeed a mould square. It is easy to see that the horizontal condition on effective fibrations implies the vertical condition of a right lifting structure of Sect. 2.2. Similarly, the perpendicular condition implies the vertical condition. It follows that the family ϕ gives p the structure of an R-

algebra. It is left to the reader to verify that this association defines a double functor between concrete double categories. □

Remark 6.3 Having defined a *double* category $\mathbb{E}$**ffFib** (from a triple category), there must be a corresponding *double* category (or cofibred structure) of arrows equipped with the left lifting property with respect to this double category, i.e. with respect to effective fibrations. These could be called *trivial cofibrations*. But note that it is not straightforward to give a presentation of this double category starting from the triple category (cf. Remark 6.2).

6.2.1 Effective Trivial Fibrations

Recall from Chap. 3 that we named the right class with respect to the double category $\mathbf{\Sigma}$ of effective cofibrations (or the dominance) *effective trivial fibrations* (Definition 3.2). Alternatively, there is a right class of arrows induced by the following triple category:

(i) Objects are the objects of $\mathcal{E}$;
(ii) Horizontal and vertical morphisms are effective cofibrations;
(iii) Perpendicular morphisms are morphisms in $\mathcal{E}$;
(iv) xy-squares are pullback squares of effective cofibrations;
(v) xz-squares are commutative squares of effective cofibrations and morphisms in $\mathcal{E}$;
(vi) yz-squares are pullback squares of an effective cofibration along a morphism in $\mathcal{E}$;
(vii) Cubes are formed by pulling back an yz square along an xz square.

For the purposes of this section, we denote the double category of arrows with a right lifting structure with respect to this triple category by $\mathbb{E}$**ffTrivFib(triple)**. In Proposition 6.1, we will see that this double category is isomorphic to $\mathbb{E}$**ffTrivFib**.

First, we observe the triple category of mould squares is a *sub*triple category of this triple category, since:

Lemma 6.3 *Every HDR is a an effective cofibration.*

Proof This is straightforward using Lemma 4.1, under the ruling assumption that every $r_X : X \to MX$ is an effective cofibration. □

In the next proposition, we have adopted the familiar notation for effective trivial fibrations. The proposition is stated in order to apply Proposition 2.7 directly.

Proposition 6.1 *There is a double functor between discretely fibred concrete double categories over* $\mathcal{E}$

$$\mathbb{E}\mathbf{ffTrivFib} \to \mathbb{E}\mathbf{ffTrivFib(triple)}$$

which is full on squares which induces an isomorphism between notions of fibred structure. Hence, this functor is an isomorphism.

Proof The first goal is to show that every effective trivial fibration can be equipped with right lifting structure with respect to the above triple category in a functorial way. Clearly, such a functor would be full on squares, since every morphism between such right lifting structures is a morphism between effective trivial fibrations. It then remains to show that the induced natural transformation between notions of fibred structure is a natural isomorphism.

Suppose $p : Y \to X$ is an effective trivial fibration. Every lifting problem with respect to an xy-square of effective cofibrations factors through a coproduct as follows:

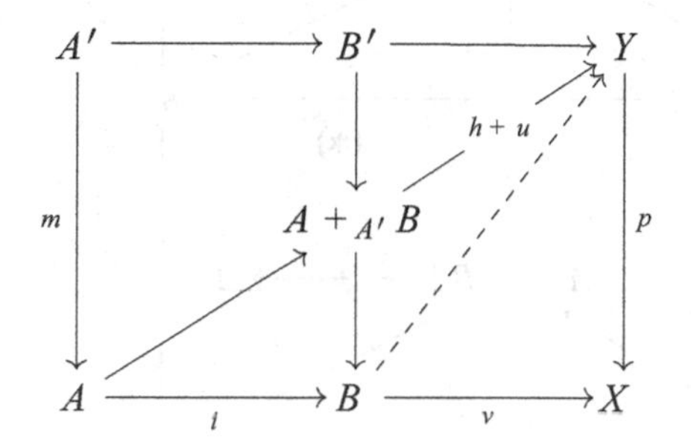

Observe that $A +_{A'} B' \hookrightarrow B$ is an effective cofibration under the prevailing assumption that these are closed under binary unions, xy squares are pullback squares and the fact that $\mathcal{E}$ is *coherent* (see Lemma 2.2). Hence there exists a filler $B \to Y$ as drawn. We need to check that it satisfies the horizontal, vertical and perpendicular conditions. For the horizontal condition, the lifting problem factors as follows:

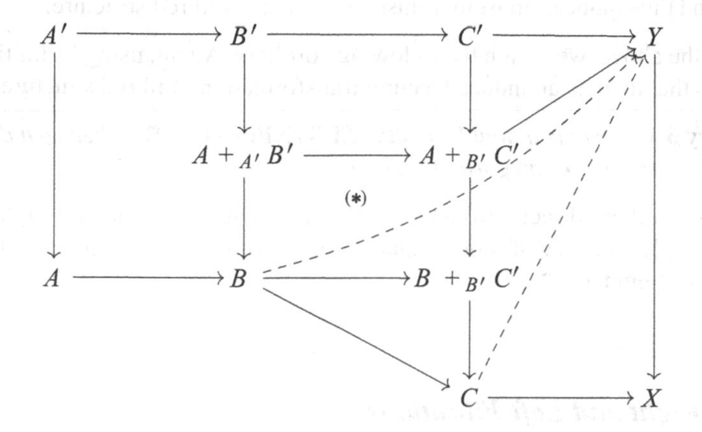

where we observe that the square $(*)$ is both a pullback and a pushout. Hence the horizontal condition for trivial fibration applies, and the lift $B \to Y$ is determined by the lift $B +_{B'} C' \to Y$. By the vertical condition, the latter is in turn compatible with the subsequent lift $C \to Y$ and equal to the lift with respect to $A +_{B'} C' \hookrightarrow C$. It follows that the dashed fillers make the whole diagram commute, which proves the horizontal condition.

The vertical condition can be proven similarly:

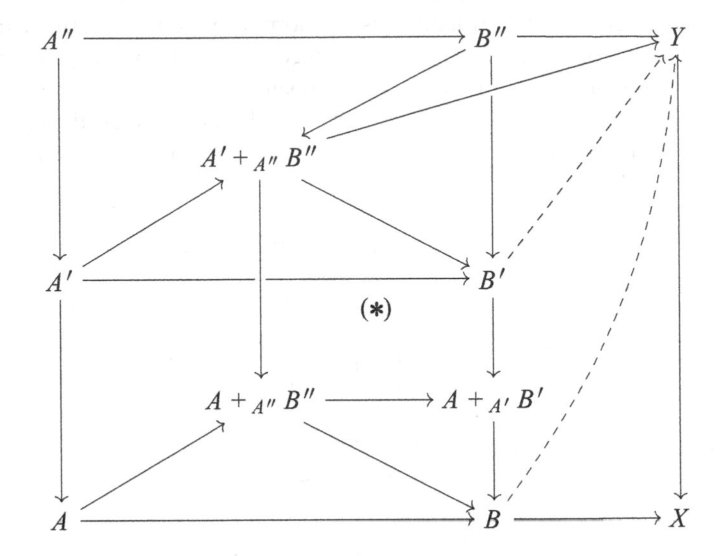

Again, the square $(*)$ is both pullback and pushout. The argument is now the same as in the horizontal case.

The perpendicular case is easy and follows from the horizontal condition for effective trivial fibrations.

It is left to the reader to convince themselves that the defined lifting structure is unique and thus induces an isomorphism of notions of fibred structure. □

From the above, we obtain the following corollary. Again, using Definition 2.14, it follows that there is an induced natural transformation of fibred structure:

Corollary 6.1 *There is a double functor* $\mathbb{E}$**ffTrivFib** $\to$ $\mathbb{E}$**ffFib** *between discretely fibred concrete double categories (over* $\mathcal{E}$*).*

Proof This follows directly from the existence of a unique (concrete) triple functor from the triple category of mould squares to the triple category in this subsection, by virtue of Lemma 6.3. □

6.2.2 Right and Left Fibrations

Having established the definition of effective fibration, we turn to the situations of Chap. 5. There, we studied the AWFS for two-sided and symmetric Moore structures. Naturally, the two-sided setting admits a dual notion of fibred structure which we call *effective left fibration* (for the terminology, recall Remark 5.1). In the symmetric case, we obtain:

Corollary 6.2 *Suppose the Moore structure on $\mathcal{E}$ is symmetric as in as in Chap. 5. Then there is a double functor between discretely fibred concrete double categories over $\mathcal{E}$*

$$\mathbb{E}\mathbf{ffFib} \to \mathbb{E}\mathbf{ffLFib}$$

which is full on squares which induces an isomorphism between notions of fibred structure. Hence, this functor is an isomorphism.

Proof This essentially follows from Lemma 6.2 and Proposition 5.1, but restricting to effective (left) fibrations. □

Chapter 7
Π-Types

This chapter contains the main result of the first part of this book. Described in more familiar terms, the result gives a constructive proof of the fact that when X is effectively fibrant, and A is any other object, then the exponential X^A is effectively fibrant (see Remark 7.1). The statement of the proposition is a more general, fibred version of this fact, which is a classic result on Kan fibrations in simplicial sets (see e.g. [36], Theorem 7.8).

We assume that $\mathcal{E}$ is a finitely cocomplete, locally cartesian closed category equipped with symmetric Moore structure and a dominance. Further, it satisfies the additional conditions from the beginning of Chap. 6. We then have the following theorem.

Theorem 7.1 (Pushforward for Effective Fibrations) *Suppose $\mathcal{E}$ is given as above. Then the pushforward $\Pi_g f$ of an effective fibration f along an effective fibration g is again an effective fibration. More precisely, there is a functor*

$$\Pi_{(-)}(-) : \mathbf{EffFib} \times_{\mathcal{E}} \mathbf{EffFib} \to \mathbf{EffFib}$$

as in Lemma 7.1 such that $\Pi_g(-)$ is right adjoint to pullback along g.

Proof By precomposing the functor of Lemma 7.1 below with the equivalence in Corollary 6.2 and the functor in Lemma 6.2, the proof follows. □

The proof refers to the following lemma, which is stated in the generality of a two-sided Moore structure on the category $\mathcal{E}$ (see the introduction to Chap. 5). Apart from that, $\mathcal{E}$ satisfies the same assumptions as in the previous theorem.

Lemma 7.1 *If $f : Y \to X$ is a naive left fibration and $g : Z \to Y$ is an effective fibration, then the pushforward $\Pi_f(g)$ is also an effective fibration. More precisely, the pullback along a naive left fibration $f^* : \mathbf{EffFib}_X \to \mathbf{EffFib}_Y$ which takes effective fibrations with codomain X to effective fibrations with codomain Y has a*

B. van den Berg, E. Faber, *Effective Kan Fibrations in Simplicial Sets*,
Lecture Notes in Mathematics 2321, https://doi.org/10.1007/978-3-031-18900-5_7

right adjoint $f_* : \mathbf{EffFib}_Y \to \mathbf{EffFib}_X$ *defined by a functor*

$$\Pi_{(-)}(-) : \mathbf{NLFib} \times_{\mathcal{E}} \mathbf{EffFib} \to \mathbf{EffFib}$$

where the domain is the pullback of the domain and codomain functors.

Proof Assume f is a naive left fibration and $g : Z \to Y$ is an effective fibration. We have to show that $\Pi_f(g)$ is an effective fibration. So imagine we have a situation like this:

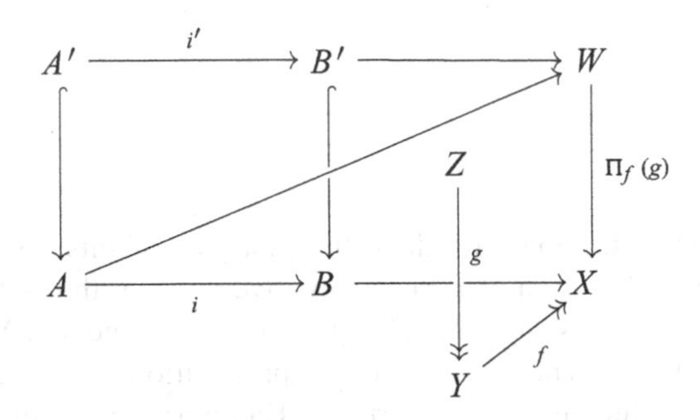

in which the left hand square is a mould square. The construction starts by taking the pullback of f along $A \to X$, which yields a naive left fibration, and subsequently constructing a 'Frobenius' cube like in (5.3):

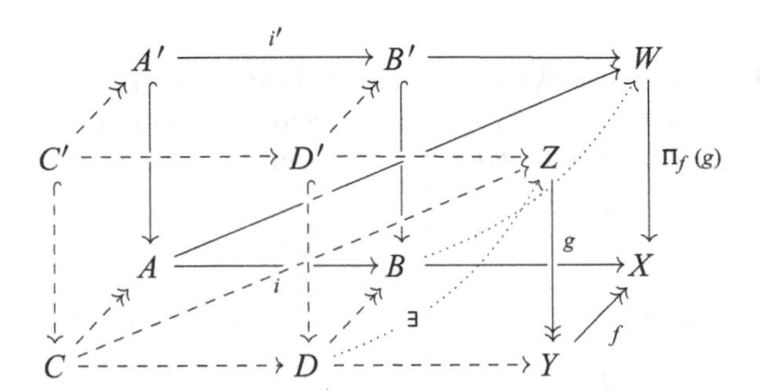

So the front and back squares are mould squares and the bottom and top are Frobenius morphisms of HDRs. The maps $D' \to Z$ and $C \to Z$ are induced by the adjunction $f^* \dashv \Pi_f$. Since g is an effective fibration, there is a map $D \to Z$ making everything commute, which we can transpose back to a map $B \to W$ giving a filler.

It remains to check the three compatibility conditions, horizontal, vertical and perpendicular, for this filler. The horizontal and vertical conditions follow fairly directly from the naturality condition for adjunctions together with Lemma 5.2.

Indeed, for the horizontal condition:

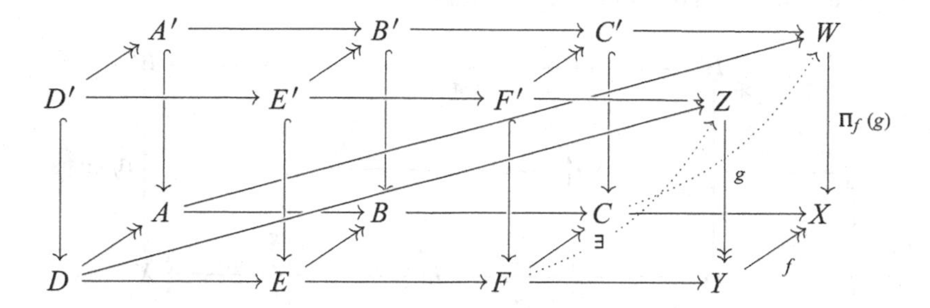

To find a map $C \to W$ we can either transpose $D \to Z$ and then push forward in one go as $F \to Z$ and then transpose back. This should coincide with: first obtaining $E \to Z$, transpose, then transpose back and compute $F \to Z$ and then transpose. It is clear that this will be the same as the other procedure.

The vertical condition is similar:

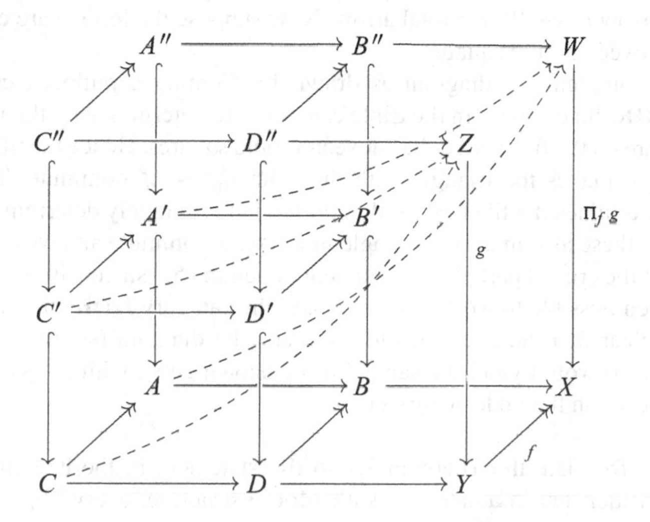

Given a map $A \to W$ one can transpose to $C \to Z$, then push forward to $D \to W$ and then transpose back. Using the vertical condition on g, this can also be done in two steps. Namely, by first lifting the restricted arrow $C' \to Z$ to $D' \to Z$, then transposing, and repeating the construction for the bottom cube. This works because the square $D'B'BD$ is a pullback, hence we return to the same diagram (same arrows) in the two-step version.

For the perpendicular condition, it helps to reduce dimensions by one by drawing HDRs as a point, as in the following diagram:

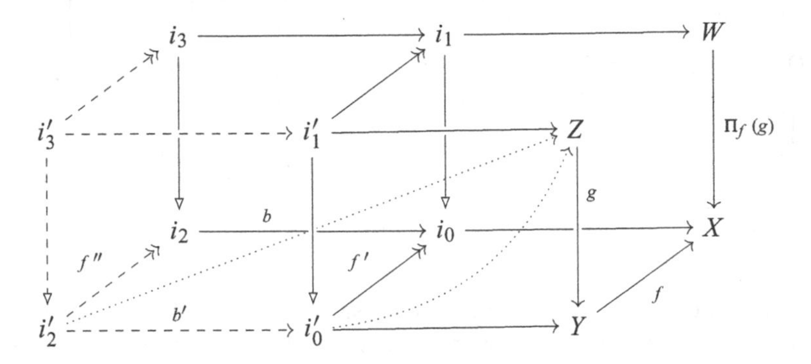

Here the open arrowtips indicate mould squares, and the double arrowtips indicate Frobenius morphisms of HDRs. In this picture, lifting can be viewed as completing a 'partial' arrow $i_0 \to W$ to a total arrow. Now suppose the left square on the back is a cube between mould squares.

We can complete the diagram as drawn by forming a pullback cube in the category **HDR**. Since cubes in the triple category are determined by their boundary, the left square on the front is a cube between mould squares. Hence the filler $i'_2 \to Z$ induced by g makes the triangle with the filler $i'_0 \to Z$ commute. The former uniquely determines the filler $i_2 \to W$, and the latter uniquely determines the filler $i_0 \to W$. So these also make the triangle in the back commute and we are done.

Note that the crucial part of this argument is Lemma 5.3 since without it, it would not have been possible to work entirely inside the category **HDR**, and it would not have been clear that the two ways to construct the diagram (starting with f' vs. starting with b) would yield the same 4-dimensional cube, with the same relevant HDR structures, in the underlying category. □

Remark 7.1 To relate the Theorem 7.1 to the statement in the beginning of this chapter, consider that in a category with Moore structure, every object is naively fibrant. Indeed, any terminal arrow $A \to 1$ can be trivially equipped with a transport structure $A \times M1 \to A$ given by projection. So every $A \to 1$ is a naive left fibration, and hence

$$X^A := \Pi_{A\to 1} X \times A \to 1$$

is an effective fibration whenever $p_1 : X \times A \to X$ is.

With the proof of Theorem 7.1, we have concluded the first part of this book. In Part II, we will show that the Moore path functor defined for simplicial sets in [19] satisfies the conditions of a (symmetric) Moore structure, and that simplicial sets admit a dominance of locally decidable monomorphisms. As a result, we obtain a notion of effective fibration in simplicial sets, which is closed under pushforward. In Chap. 12, we will show that effective fibrations in simplicial sets are always Kan fibrations, and that under classical assumptions, the notions coincide.

Part II
Simplicial Sets

Chapter 8
Effective Trivial Kan Fibrations in Simplicial Sets

Whereas in the first part we derived the existence of Π-types in an axiomatic setting based on a suitable combination of a dominance and a symmetric Moore structure, this second part will be entirely devoted to one particular example: simplicial sets. To show that the category of simplicial sets is indeed an example, we will first have to choose both a dominance and a symmetric Moore structure on simplicial sets, and the first two chapters of this second part will do exactly that. Indeed, in this chapter we will choose a dominance and in the next chapter we will choose a symmetric Moore structure. After that, we will study the resulting HDRs and effective Kan fibrations in more detail. In particular, we will show that the effective Kan fibrations are a local notion of fibred structure and that in a classical metatheory the maps which can be equipped with the structure of an effective Kan fibration are precisely those maps which have the right lifting property against horn inclusions.

But before we get into that, let us first choose and study a suitable dominance on simplicial sets. As we have seen in Chap. 3, dominances induce AWFSs. We will call the coalgebras for the comonad of the induced AWFS *effective cofibrations*, while the algebras for the monad will be called the *effective trivial Kan fibrations*. The main results of this chapter will be that being an effective trivial Kan fibration is a local notion fibred structure, and that (in a classical metatheory) a map can be equipped with structure of an effective trivial Kan fibration if and only if it has the right lifting property against boundary inclusions.

8.1 Effective Cofibrations

Traditionally, the cofibrations in simplicial sets are simply the monomorphisms. In our constructive metatheory we believe it is important to add a decidability condition. We will further comment on this choice at the end of this section.

B. van den Berg, E. Faber, *Effective Kan Fibrations in Simplicial Sets*,
Lecture Notes in Mathematics 2321, https://doi.org/10.1007/978-3-031-18900-5_8

Definition 8.1 In the category of simplicial sets we will call $m : B \to A$ an *(effective) cofibration* if it is a locally decidable monomorphism : that is, each $m_n : B_n \to A_n$ is a complemented monomorphism in the subobject lattice of A_n. In other words, for each $a \in A_n$ we can decide whether there is an element $b \in B_n$ such that $m_n(b) = a$ or not.

In the statement of the following lemma we will say that a set is finite if it can be put in bijective correspondence with some initial segment of the natural numbers. Moreover, we will say that a sieve S is generated by a set of maps I if S is the closure of I under precomposition with arbitrary maps.

Lemma 8.1 *The following are equivalent for a sieve $S \subseteq \Delta^n$:*

(1) The inclusion $i : S \subseteq \Delta^n$ is an effective cofibration.
(2) The sieve is generated by a finite set of monos $\Delta^m \to \Delta^n$.
(3) The sieve is generated by a finite set of maps.

Proof The implications (2) ⇒ (3) ⇒ (1) are obvious, so it remains to show that (1) ⇒ (2).

But since every map in $\mathbf{\Delta}$ factors as an epi followed by a mono (in a unique way), and every epi splits, every sieve is generated by its monomorphisms. But since there are only finitely many monos with codomain Δ^n, a cofibrant sieve contains only finitely many monos. □

Definition 8.2 We will refer to the sieves satisfying any of the equivalent conditions in the previous lemma as the *cofibrant sieves*.

Because a monomorphism in $\mathbf{\Delta}$ with codomain $[n]$ is completely determined by its image, a cofibrant sieve can be thought of as a subsimplicial complex of Δ^n, that is, a collection of inhabited and decidable subsets of $\{0, 1, \ldots, n\}$ which is itself closed under inhabited and decidable subsets.

Theorem 8.1 *The effective cofibrations in simplicial sets form a dominance.*

Proof The cofibrations are clearly closed under pullback and composition, so we only need to prove that there is a cofibration $1 \to \Sigma$ such that any other can be obtained as a pullback of that one in a unique way. We put

$$\Sigma_n := \{ S \subseteq \Delta^n \ : \ S \text{ cofibrant sieve} \}.$$

(Note that Σ_n is finite, so that Σ_n is a set even in a predicative metatheory like **CZF**.) Since cofibrant sieves are stable under pullback along $\alpha : \Delta^m \to \Delta^n$, this defines a simplicial set. That is, we define the action on Σ by the following formula:

$$S \cdot \alpha = \{\beta : [k] \to [m] \ : \ \alpha.\beta \in S \}.$$

In addition, there is a natural transformation $\top : 1 \to \Sigma$ obtained by picking the maximal sieve at each level. This map classifies the cofibrations in that for any cofibration $m : B \to A$ the map $\mu : A \to \Sigma$ defined by

$$\mu_n(a) = \{\alpha : [m] \to [n] \, : \, (\exists b \in B_m)\, a \cdot \alpha = m(b)\, \}$$

turns

$$\begin{array}{ccc} B & \longrightarrow & 1 \\ {\scriptstyle m}\downarrow & & \downarrow{\scriptstyle \top} \\ A & \underset{\mu}{\longrightarrow} & \Sigma \end{array}$$

into a pullback. Also, the map μ is easily seen to be unique with this property. □

Remark 8.1 Note that the cofibrant subobjects form a sub-Heyting algebra of the full subobject lattice. In particular, the cofibrant sieves are closed under all the propositional operations: not only $\wedge, \top$, but also $\bot, \vee$ and $\to$. To see that they are closed under implication, for instance, note that for sieves $S, T \subseteq \Delta^n$, we have

$$\alpha : [m] \to [n] \in (S \to T) \iff (\forall \beta : [k] \to [m])\, (\alpha.\beta \in S \Rightarrow \alpha.\beta \in T\,).$$

Because maps in $\mathbf{\Delta}$ factor as an epi followed by a mono, and epis split, we only need to check the condition on the right for monos β. So if S and T are cofibrant, the condition on the right is decidable and $S \to T$ is cofibrant as well.

8.2 Effective Trivial Kan Fibrations

Since the effective cofibrations in simplicial sets form a dominance, they are the left class in an algebraic weak factorisation system. The members of the right class will be referred to as the *effective trivial Kan fibrations*. From the work by Bourke and Garner (recapitulated in Chap. 2), we know that these can be defined as the maps which come with a compatible system of lifts against a large double category: one where the vertical maps are the cofibrations and the squares are pullback squares. Our first goal in this section is to show that we can restrict attention to a small subdouble category.

Indeed, let $\mathbb{C}$ be the following small double category:

- Objects are cofibrant sieves $S \subseteq \Delta^n$.
- Horizontal maps from $S \subseteq \Delta^n$ to $T \subseteq \Delta^m$ are maps $\alpha : \Delta^n \to \Delta^m$ such that $T \cdot \alpha = S$.
- Vertical maps are inclusions of cofibrant sieves $S_0 \subseteq S_1 \subseteq \Delta^n$.

- Squares are pullback diagrams of the form

$$\begin{array}{ccc} S_0 & \xrightarrow{\alpha} & T_0 \\ \downarrow & & \downarrow \\ S_1 & \xrightarrow[\alpha]{} & T_1 \end{array}$$

such that both horizontal maps are labelled with the same α.

Clearly, there is an inclusion of double categories from $\mathbb{C}$ to the large double category of cofibrations.

Proposition 8.1 *The following notions of fibred structure are isomorphic:*

- *Having the right lifting property against the large double category of cofibrations (that is, to be an effective trivial Kan fibration).*
- *Having the right lifting property against the small double category* $\mathbb{C}$.

In fact, if Cof *is the large double category of cofibrations, then the morphism of discretely fibred concrete double categories* $\mathrm{Cof}^{\pitchfork} \to \mathbb{C}^{\pitchfork}$ *induced by the inclusion is an isomorphism.*

Proof Assume $p : Y \to X$ has the right lifting property against the small double category $\mathbb{C}$, and imagine that we have a lifting problem of the form:

$$\begin{array}{ccc} B & \longrightarrow & Y \\ {\scriptstyle m}\downarrow & \overset{l}{\nearrow} & \downarrow{\scriptstyle p} \\ A & \longrightarrow & X. \end{array}$$

Suppose $a \in A_n$ is arbitrary and we pull back m along $a : \Delta^n \to A$:

$$\begin{array}{ccccc} S & \longrightarrow & B & \longrightarrow & Y \\ \downarrow & & \downarrow{\scriptstyle m} & & \downarrow{\scriptstyle p} \\ \Delta^n & \xrightarrow[a]{} & A & \longrightarrow & X. \end{array} \tag{8.1}$$

Since the pullback is a cofibrant sieve, we find an element $y \in Y_n$ filling the outer rectangle, and we put $l_n(a) := y$. Note that this definition is forced, because the left hand square in the diagram 8.1 is a square in the large double category. Note also that l is going to be a natural transformation because of the horizontal condition coming from the pullback squares in $\mathbb{C}$:

$$\begin{array}{ccccccc} S\cdot\alpha & \longrightarrow & S & \longrightarrow & B & \longrightarrow & Y \\ \downarrow & & \downarrow & & \downarrow{\scriptstyle m} & & \downarrow{\scriptstyle p} \\ \Delta^m & \xrightarrow[\alpha]{} & \Delta^n & \xrightarrow[a]{} & A & \longrightarrow & X. \end{array}$$

(Here the words horizontal and vertical condition refer to the conditions for being a right lifting structure, as can be found after Example 2.2.)

Next, let us check that in case m is a vertical map coming from the small double category, the new lift l agrees with the one coming from the fact that p has the right lifting property against $\mathbb{C}$: for note that if $S \subseteq T \subseteq \Delta^n$ are cofibrant sieves and $\alpha : \Delta^m \to \Delta^n \in T$, then $T \cdot \alpha = \Delta^m$ and the left hand square in

$$\begin{array}{ccccc} S \cdot \alpha & \longrightarrow & S & \longrightarrow & Y \\ \downarrow & y & \downarrow & & \downarrow p \\ \Delta^m & \longrightarrow & T & \longrightarrow & X \end{array}$$

is a square in the double category $\mathbb{C}$. So both the lift $T \to Y$ we have constructed and the one coming from the fact that p has the right lifting property against $\mathbb{C}$ send α to the lift y.

It is now easy to see that the constructed lifts satisfy both the horizontal and vertical conditions with respect to the large double category, thus showing that on the level of notions of fibred structure we have an isomorphism. The fullness condition for squares follows from the fact that the left hand square in diagram 8.1 belongs to the large double category, showing that we have an isomorphism of discretely fibred concrete double categories. □

We will now cut down things even further. In fact, the lifting structure against $\mathbb{C}$ is completely determined by its lifts against the boundary inclusions, as we will now show.

Lemma 8.2 *Suppose $p : Y \to X$ has two lifting structures against the small double category $\mathbb{C}$. If these two lifting structures agree on the lifts against the boundary inclusions, then they agree on all vertical maps.*

Proof Let $S \subseteq T \subseteq \Delta^n$ be cofibrant sieves. Then this inclusion can be decomposed as

$$S = S_0 \subseteq S_1 \subseteq S_2 \subseteq \ldots \subseteq S_k = T \subseteq \Delta^n$$

where each S_{i+1} contains precisely one face of Δ^n more than S_i (for $0 \leq i < k$). By the vertical condition, the lift against $S \subseteq T$ is completely determined by the lifts against the $S_i \subseteq S_{i+1}$. But if $\Delta^m \to S_{i+1}$ is the one face which belongs to S_{i+1} but not S_i, then its entire boundary lies in S_i and we have a pullback diagram as follows:

$$\begin{array}{ccc} \partial\Delta^m & \longrightarrow & S_i \\ \downarrow & & \downarrow \\ \Delta^m & \longrightarrow & S_{i+1}. \end{array}$$

Since this diagram is both a square in the double category $\mathbb{C}$ and a pushout in simplicial sets, the lift against the map on the right is completely determined by the lift against the map on the left. □

In the remainder of this section we will try to answer the following question: suppose we have a map $p : Y \to X$ and we have chosen lifts against all boundary inclusions

$$\begin{array}{ccc} \partial\Delta^m & \longrightarrow & Y \\ \downarrow & \overset{f_i}{\nearrow} & \downarrow p \\ \Delta^m & \longrightarrow & X. \end{array}$$

What conditions do these lifts f_i have to satisfy in order for them to extend to a lifting structure against $\mathbb{C}$?

First of all, because any inclusion of sieves $S \subseteq T \subseteq \Delta^n$ can be seen as a composition of pushouts of boundary inclusions, as in the previous lemma, we can solve any lifting problem of the form

$$\begin{array}{ccc} S & \longrightarrow & Y \\ \downarrow & \nearrow & \downarrow p \\ T & \longrightarrow & X. \end{array}$$

The first worry is that the decomposition of the inclusion $S \subseteq T$ as a sequence of inclusions where the next sieve contains one face more than the previous is in no way unique, and it could be that the lift we construct depends on the decomposition. As a matter of fact, it does not depend on this: imagine that we choose two different decompositions of the inclusion $S \subseteq T$ and they determine lifts f and g, respectively. Now we can prove by induction on $n \in \mathbb{N}$ that f and g agree on all n-simplices, using that they agree on their boundaries in the induction step.

The next worry is that these lifts need to satisfy both the horizontal and the vertical condition coming from $\mathbb{C}$. The vertical condition is, in fact, automatically satisfied, because of the way we constructed the lifts and the fact that the way we decompose the vertical maps in $\mathbb{C}$ is irrelevant.

Therefore we only need to consider the horizontal condition: that is, we need to determine which conditions the chosen lifts f_i need to satisfy in order that in any diagram of the form

$$\begin{array}{ccccc} \alpha^*S & \longrightarrow & S & \longrightarrow & Y \\ \downarrow & & \downarrow & \nearrow & \downarrow p \\ \alpha^*T & \longrightarrow & T & \longrightarrow & X \end{array}$$

with the square on the left a square in $\mathbb{C}$, the induced lifts make the resulting triangle commute. The horizontal condition can be split in two: because every map in $\mathbf{\Delta}$ is the composition of face and degeneracy maps, we only need to worry about squares where the map α is either a face or degeneracy map. In fact, the case where α is a face map is unproblematic. The reason is this: imagine that we have a decomposition

$$S_0 \subseteq S_1 \subseteq S_2 \subseteq \ldots \subseteq S_k \subseteq \Delta^n$$

of $S \subseteq T$ and each S_{j+1} contains precisely one face more than S_j. If we pull this back along $d_i : \Delta^{n-1} \to \Delta^n$, we get either that $d_i^* S_j = d_i^* S_{j+1}$ if the face that gets added to S_j in this step does not belong to d_i, or that $d_i^* S_j \neq d_i^* S_{j+1}$ in case the face that gets added to S_j in this step does belong to d_i. But in the latter case, $d_i^* S_{j+1}$ contains one face more than $d_i^* S_j$, so if we ignore all the first cases we obtain a decomposition of $d_i^* S_0 \subseteq d_i^* S_k$. If we use this decomposition to compute the lift against $d_i^* S_0 \subseteq d_i^* S_k$, then by pullback pasting

$$\begin{array}{ccccc} \partial\Delta^m & \longrightarrow & d_i^* S_j & \longrightarrow & S_j \\ \downarrow & & \downarrow & & \downarrow \\ \Delta^m & \longrightarrow & d_i^* S_{j+1} & \longrightarrow & S_j \end{array}$$

it is computed in exactly the same way as the lift against $S_0 \subseteq S_k$ is computed on the simplices which belong to the ith face.

So the upshot of the discussion so far is that we only need to worry about the horizontal condition for squares with α being a degeneracy map. Here, in view of the decomposition, we can restrict attention to the situation where the map on the right in the square is an inclusion $S \subseteq T$ where T contains precisely one face more than S. In fact, we claim that we only need to worry about the situation where the map on the right is a boundary inclusion, as in:

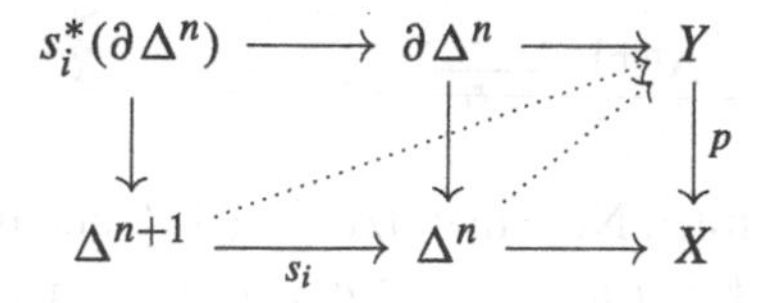

Indeed, assume that $S \subseteq T \subseteq \Delta^n$ is an inclusion where T contains precisely one more face than S, which happens to be $\Delta^m \to T$; also assume that we have some lifting problem of $S \subseteq T$ against p, and $s_i : \Delta^{n+1} \to \Delta^n$ is some degeneracy. Using that pullbacks of monos along epis exist in the simplicial category, we can create a diagram as follows:

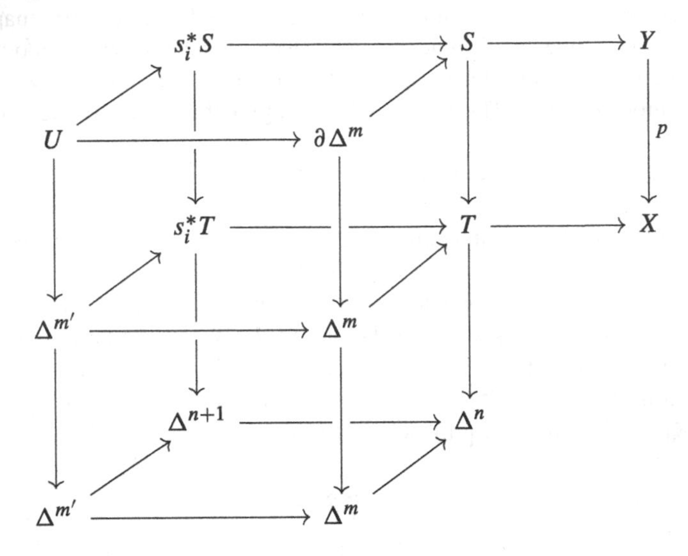

Note that in the top cube, the left, right, front and back faces are squares in the double category $\mathbb{C}$, and the right face is a pushout of simplicial sets. Therefore the left face is a pushout as well and the horizontal condition for the back face of that cube is equivalent to the horizontal condition for the front of that cube. But the map $\Delta^{m'} \to \Delta^m$ is either the identity if i does not belong to the image of $\Delta^m \to \Delta^n$, or some degeneracy $s_j : \Delta^{m+1} \to \Delta^m$ if it does.

This means that we can restrict attention to the following situation:

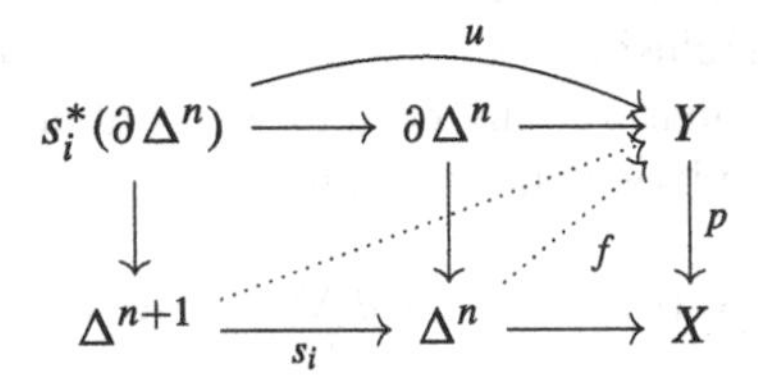

and let f be our chosen filler. Note that $\partial\Delta^n = \bigcup d_k^n$ and the pullback of d_k along s_i is d_k if $k < i$, $d_i.d_i$ if $k = i$ and d_{k+1} if $k > i$; in other words, $s_i^*(\partial\Delta^n)$ is Δ^{n+1} with the interior and ith and $(i + 1)$st faces missing. So to find the dotted filler in the diagram above we first need to find the filler on the faces i and $i + 1$. So we pull back the left hand arrow along d_i and d_{i+1} and choose our chosen filler, which is, actually, f, because $d_i.s_i = d_{i+1}.s_i = 1$. So we are left with the following filling problem:

$$\begin{array}{ccc} \partial\Delta^{n+1} & \xrightarrow{u \cup f \cup f} & Y \\ \downarrow & \nearrow & \downarrow p \\ \Delta^{n+1} & \longrightarrow & X. \end{array}$$

So what we need is that the chosen solution of this problem will be $f \cdot s_i$. (Note that $f \cdot s_i$ will always be a solution. Indeed, we have $f \cdot s_i \cdot d_k = (u \cup f \cup f) \cdot d_k$ for any face d_k: it is true on the faces that we added (d_i and d_{i+1}), but also on the faces that were already there, because the original picture commutes.)

So, to summarise our discussion, we have:

Theorem 8.2 *The following notions of fibred structure are naturally isomorphic:*

- *To assign to each $p : Y \to X$ all effective trivial Kan fibration structures on it.*
- *To assign to each $p : Y \to X$ all systems of lifts of p against boundary inclusions such that if f is our chosen filler of*

$$\begin{array}{ccc} \partial\Delta^{n} & \xrightarrow{y} & Y \\ \downarrow & \overset{f}{\nearrow} & \downarrow p \\ \Delta^{n} & \xrightarrow[x]{} & X, \end{array}$$

then $f.s_i$ is our chosen solution of the problem

$$\begin{array}{ccc} \partial\Delta^{n+1} & \xrightarrow{y'} & Y \\ \downarrow & \nearrow & \downarrow p \\ \Delta^{n+1} & \xrightarrow[x.s_i]{} & X, \end{array}$$

where y' is the composition $s_i^*(\partial\Delta^n) \to \partial\Delta^n \to Y$ on $s_i^*(\partial\Delta^n)$ and f on d_i and d_{i+1}.

Remark 8.2 Clearly, both the effective trivial fibrations and the maps which carry a lifting structure against all boundary inclusions satisfying the condition in the theorem above are the vertical maps in concrete double categories. The statement of the theorem can be strengthened to say that these concrete double categories are isomorphic. Indeed, it suffices to check the fullness condition for squares for the obvious forgetful functor of double categories: but that can be shown as in Lemma 8.2.

8.3 Local Character and Classical Correctness

The characterisation given in Theorem 8.2 can both be used to show that our notion of an effective trivial Kan fibration is local and that it is classically correct. Let us first discuss local character.

Corollary 8.1 *The notion of an effective trivial Kan fibration is a local notion of fibred structure.*

Proof As said, we use the characterisation in Theorem 8.2. So assume $p : Y \to X$ is a map of simplicial sets such that any pullback of it along a map $x : \Delta^n \to X$ is an effective trivial Kan fibration. If we have a lifting problem of the form

$$\begin{array}{ccc} \partial\Delta^n & \xrightarrow{y} & Y \\ \downarrow & & \downarrow{\scriptstyle p} \\ \Delta^n & \xrightarrow[x]{} & X, \end{array}$$

then we can decompose it as follows

$$\begin{array}{ccccc} \partial\Delta^n & \xrightarrow{y} & Y_x & \longrightarrow & Y \\ \downarrow & & \downarrow & & \downarrow{\scriptstyle p} \\ \Delta^n & \xrightarrow[1]{} & \Delta^n & \xrightarrow[x]{} & X, \end{array}$$

with a pullback on the right. Since the lifting problem on the right has a solution, by assumption, we also get a filler f for the outer rectangle. Since this definition is forced, we only need to check the condition for the degeneracies. So then we are looking at a situation like this:

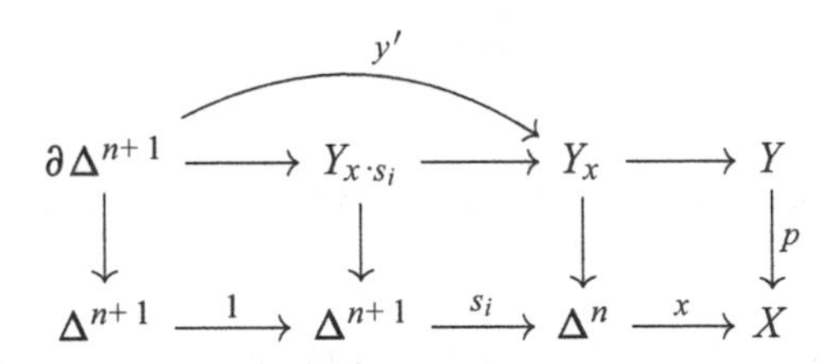

The lift against p is induced by the left hand square, but, by assumption, it is compatible with the one coming from the composition of the two squares on the left, which is $f.s_i$, as desired. □

It remains to check classical correctness, for which we use the following lemma, whose proof can be found in the appendix (see Proposition C.1).

Lemma 8.3 *A lifting problem of the form*

$$\begin{array}{ccc} \partial\Delta^n & \longrightarrow & X \\ \downarrow & \nearrow & \\ \Delta^n & & \end{array}$$

has at most one degenerate solution (that is, if both $x_0 \cdot \sigma_0$ *and* $x_1 \cdot \sigma_1$ *fill this triangle with both* σ_i *epis in* $\mathbf{\Delta}$ *different from the identity, then* $x_0 \cdot \sigma_0 = x_1 \cdot \sigma_1$*).*

Theorem 8.3 *Classically, any morphism which has the right lifting property with respect to boundary inclusions* $\partial\Delta^n \subseteq \Delta^n$ *can be equipped with the structure of an effective trivial Kan fibration.*

Proof Suppose $p : Y \to X$ is a map for which we have a choice of fillers f_i for every lifting problem of the form

$$\begin{array}{ccc} \partial\Delta^n & \longrightarrow & Y \\ \downarrow & \overset{f_i}{\nearrow} & \downarrow p \\ \Delta^n & \longrightarrow & X. \end{array}$$

(This uses the axiom of choice, depending on how one reads the assumption.) Then, using the Principle of Excluded Middle and the previous lemma, we may assume that f_i is the unique degenerate solution, if it exists. Then the compatibility condition from Theorem 8.2 is automatically satisfied, because it says that under certain conditions we should choose the (unique) degenerate solution. □

Remark 8.3 The fact that cofibrations are pointwise complemented plays an important role in the work on cubical sets in the proof of the univalence axiom (the axiomatic treatment in [20] makes this quite clear). We expect that we will need this assumption for similar reasons and that is our main reason for including it. An additional benefit is that this ensures that, even in a predicative metatheory like **CZF**, the effective cofibrations are a dominance and the effective trivial Kan fibrations are cofibrantly generated by a small double category, as we have seen in this chapter.

In their work Gambino, Henry, Sattler and Szumiło [15–17] work with a stronger notion of cofibration, where the question whether an n-simplex is degenerate or not is decidable for those n-simplices outside the image of the cofibration. For us this choice would have several unfortunate consequences: one consequence will be that HDRs will not be cofibrations and effective trivial Kan fibrations might not be effective Kan fibrations (see Corollary 6.1 and Proposition 10.1 below). Additionally, it will have the consequence that not every simplicial set is cofibrant.

On the other hand, one may wonder whether the previous result can be made more constructive when degeneracy is decidable (in Y for instance). We fail to see how it would, and for that reason the relationship with the work of Gambino et al. is far from clear to us.

Chapter 9
Simplicial Sets as a Symmetric Moore Category

The purpose of this chapter is to show that the category of simplicial sets can be equipped with symmetric Moore structure. As we already mentioned in the introduction, the structure that we choose was already defined in the paper by Van den Berg and Garner [19] using the notion of simplicial Moore paths. However, since the notion of a symmetric Moore structure that we work with in this book is stronger than that of a path object category as in [19], we have to verify some additional equations. For checking that these hold, we use a new characterisation of the Moore path functor M, namely, as a polynomial functor. For that reason we will first give a brief introduction to the theory of polynomial functors. Then we will define M as a polynomial functor and use this definition to check that it is a symmetric Moore structure in the sense of this book. Finally, we will prove that our new definition of M is equivalent to the one given in [19]. We will also isolate an interesting two-sided Moore structure on simplicial sets, which will give us effective left and right fibrations.

Incidentally, we would like to point out that the simplicial Moore path functor M had already occurred in the work by Clemens Berger, before it was rediscovered by Van den Berg and Garner. (It was called I in [37]; the functor called I in [38] is related, but different.)

9.1 Polynomial Yoga

We start by recapping some general facts about polynomial functors. (Some useful references are: [39–41].) Throughout this section we will work in a category $\mathcal{E}$ which is locally cartesian closed and has finite colimits. (We assume all this structure is chosen.)

Definition 9.1 A *polynomial* in $\mathcal{E}$ is a morphism $f : B \to A$ in $\mathcal{E}$. For reasons that will become clear soon, we will also write such morphisms as $(B_a)_{a \in A}$. We

B. van den Berg, E. Faber, *Effective Kan Fibrations in Simplicial Sets*,
Lecture Notes in Mathematics 2321, https://doi.org/10.1007/978-3-031-18900-5_9

will refer to A as the *base* and B_a as a *fibre*. A *morphism* α *of polynomials* from $f : B \to A$ to $g : D \to C$ is a pair (α^+, α^-) consisting of a morphism $\alpha^+ : A \to C$ and a morphism $\alpha^- : A \times_C D \to B$ making

$$\begin{array}{ccccc} B & \xleftarrow{\alpha^-} & A \times_C D & \longrightarrow & D \\ & \searrow^{f} & \downarrow & & \downarrow g \\ & & A & \xrightarrow{\alpha^+} & C \end{array}$$

commute. We will refer to α^+ as the *positive* or *forward direction* of the morphism α, while α^- is its *negative* or *backward direction*. So, basically, a morphism α from $(B_a)_{a \in A}$ to $(D_c)_{c \in C})$ consists of a map $\alpha^+ : A \to C$ and a family of morphisms $(\alpha_a^- : D_{\alpha^+(a)} \to B_a)_{a \in A}$. With this notation, composition of morphisms is given by

$$(\beta^+ : C \to E, \beta_c^- : F_{\gamma^+(c)} \to D_c) \circ (\alpha^+ : A \to C, \alpha_a^- : D_{\alpha^+(a)} \to B_a) =$$
$$(\beta^+.\alpha^+ : A \to E, \alpha_a^-.\beta_{\alpha^+(a)}^- : F_{\beta^+(\alpha^+(a))} \to D_{\alpha^+(a)} \to B_a).$$

The result is a category which we will denote $\mathrm{Poly}(\mathcal{E})$.

In addition, let us write $\mathrm{FEnd}(\mathcal{E})$ for the category of fibred endofunctors on $\mathcal{E}$ and fibred natural transformation between them (with respect to the codomain fibration on $\mathcal{E}$; see [42]). There is a functor $P : \mathrm{Poly}(\mathcal{E}) \to \mathrm{FEnd}(\mathcal{E})$ sending a polynomial $f : B \to A$ to its associated *polynomial functor* P_f:

$$P_f : \mathcal{E} \xrightarrow{B^*} \mathcal{E}/B \xrightarrow{\Pi_f} \mathcal{E}/A \xrightarrow{\Sigma_A} \mathcal{E}.$$

Written differently:

$$P_f(X) = \sum_{a \in A} \prod_{b \in B_a} X = \sum_{a \in A} X^{B_a} = \{(a \in A, t : B_a \to X)\}.$$

On morphisms $\alpha : (f : B \to A) \to (g : D \to C)$, the functor P acts as follows:

$$P(\alpha)_X : P_f(X) \to P_g(X) : (a \in A, t : B_a \to X) \mapsto (\alpha^+(a) \in C, t.\alpha_a^- : D_{\alpha^+(a)} \to B_a \to X).$$

Note that $P(\alpha)$ is a cartesian natural transformation (meaning: all naturality squares are pullbacks) if α^- is iso.

The following proposition will not be used in this book, but explains the choice of morphisms in the category $\mathrm{Poly}(\mathcal{E})$.

Proposition 9.1 *The functor* $P : \mathrm{Poly}(\mathcal{E}) \to \mathrm{FEnd}(\mathcal{E})$ *is full and faithful.*

Proof See [40, Theorem 3.4]. □

The category FEnd($\mathcal{E}$) has finite limits and these are inherited by Poly($\mathcal{E}$). The terminal object is the polynomial $0 \to 1$. The product of $(f : B \to A) \times (g : D \to C)$ is

$$[1_A \times f, g \times 1_B] : A \times D + B \times C \to A \times C.$$

In other words, it has $A \times C$ as base, with fibre $D_a + B_c$ over $(a, c) \in A \times C$. The pullback of $\delta : (g : D \to C) \to (f : B \to A)$ and $\varphi : (h : F \to E) \to (f : B \to A)$ has $C \times_A E$ as base, with the fibre $P_{(c,e)}$ over (c, e) being the pushout:

$$\begin{array}{ccc} B_{\delta^+(c)} = B_{\varphi^+(e)} & \xrightarrow{\delta_c^-} & D_c \\ {\scriptstyle \varphi_e^-}\downarrow & & \downarrow \\ F_e & \longrightarrow & P_{(c,e)}. \end{array}$$

In addition, the category FEnd($\mathcal{E}$) carries a (non-symmetric) monoidal structure given by composition: $F \otimes G = F \circ G$. This is inherited by Poly($\mathcal{E}$) as well: indeed, it carries a monoidal structure as follows:

$$(B_a)_{a \in A} \otimes (D_c)_{c \in C} = \{(b \in B_a, d \in D_{t(b)})\}_{(a \in A, t : B_a \to C)}.$$

Imagine that we have a morphism $\alpha : (B_a)_{a \in A} \to (B'_{a'})_{a' \in A'}$ and a morphism $\gamma : (D_c)_{c \in C} \to (D'_{c'})_{c' \in C'}$ then $\alpha \otimes \gamma = \eta$ with:

$$\eta^+(a \in A, t : B_a \to C) = (\alpha^+(a) \in A', \gamma^+.t.\alpha_a^- : B'_{\alpha^+(a)} \to B_a \to C \to C')$$

$$\eta^-_{(a \in A, t : B_a \to C)}(b' \in B'_{\alpha^+(a)}, d' \in D'_{(\gamma^+.t.\alpha_a^-)(b')}) = (\alpha_a^-(b') \in B_a, \gamma^-_{t(\alpha_a^-(b'))}(d') \in D_{t(\alpha_a^-(b'))})$$

The monoidal unit I is $1 \to 1$ (corresponding to the identity functor).

Proposition 9.2 *The tensor $\otimes$ on* Poly($\mathcal{E}$) *preserves pullbacks in both coordinates.*

Proof Because pullbacks in functor categories are computed pointwise and polynomial functors preserve pullbacks. □

We will also be interested in comonoids for this tensor: so, this consists of an object M in Poly($\mathcal{E}$) together with maps $\varepsilon : M \to I$ (the counit) and $\delta : M \to M \otimes M$ (the comultiplication) making the following

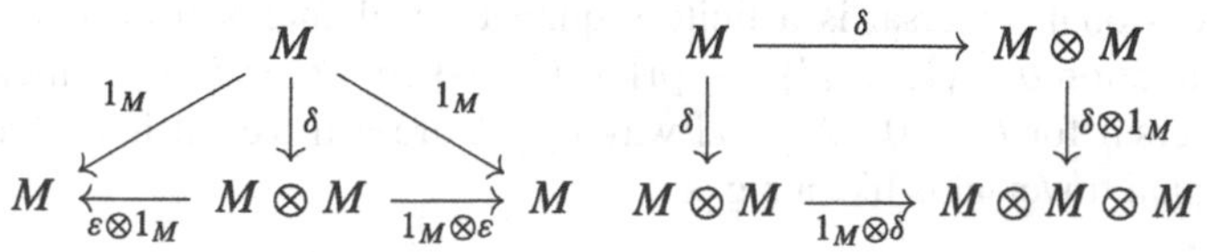

$$\begin{array}{ccccc} & & M & & \\ & \swarrow^{1_M} & \downarrow{\scriptstyle \delta} & \searrow^{1_M} & \\ M & \xleftarrow[\varepsilon \otimes 1_M]{} & M \otimes M & \xrightarrow[1_M \otimes \varepsilon]{} & M \end{array} \qquad \begin{array}{ccc} M & \xrightarrow{\delta} & M \otimes M \\ {\scriptstyle \delta}\downarrow & & \downarrow{\scriptstyle \delta \otimes 1_M} \\ M \otimes M & \xrightarrow[1_M \otimes \delta]{} & M \otimes M \otimes M \end{array}$$

commute. We will refer to such objects as *polynomials comonads*.

Every internal category in $\mathcal{E}$ determines such a comonad. Indeed, let $\mathbb{C}$ be an internal category and $\mathrm{cod} : \mathbb{C}_1 \to \mathbb{C}_0$ be the codomain map. Then there is a counit:

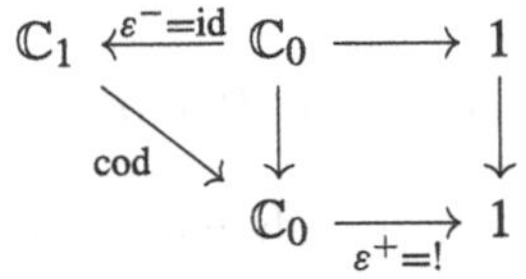

$$
\begin{array}{ccccc}
\mathbb{C}_1 & \xleftarrow{\varepsilon^-=\mathrm{id}} & \mathbb{C}_0 & \longrightarrow & 1 \\
& \searrow^{\mathrm{cod}} & \downarrow & & \downarrow \\
& & \mathbb{C}_0 & \xrightarrow[\varepsilon^+=!]{} & 1
\end{array}
$$

and a comultiplication $(\delta^+, \delta^-) : \mathrm{cod} \to \mathrm{cod} \otimes \mathrm{cod}$ with $\delta^+ : \mathbb{C}_0 \to \sum_{C \in \mathbb{C}_0} \mathbb{C}_0^{\mathrm{cod}^{-1}(C)}$ given by sending an object C in $\mathbb{C}$ to the pair $(C, \lambda\alpha \in \mathrm{cod}^{-1}(C).\mathrm{dom}(\alpha))$, whilst $(\delta^-)_C$ sends a pair $(\alpha : D \to C, \beta : E \to D)$ to $\alpha.\beta$.

This construction has a converse: indeed, one can show that every polynomial comonad is induced in this way by an internal category (see [43]).

Note that such a polynomial comonad is in particular a comonad (in the usual sense) on $\mathcal{E}$ and that the coalgebras for this comonad are precisely the internal presheaves on $\mathbb{C}$ in $\mathcal{E}$. Note also that such a polynomial comonad is automatically strong. Indeed, because a polynomial functor preserves pullbacks, we can think of a strength on P_{cod} as a natural transformation $\alpha_X : X \times P_{\mathrm{cod}}(1) \to P_{\mathrm{cod}}(X)$, or, in other words, as a map of polynomials $(1 : \mathbb{C}_0 \to \mathbb{C}_0) \cong (1 \to 1) \times (0 \to \mathbb{C}_0) \to (\mathrm{cod} : \mathbb{C}_1 \to \mathbb{C}_0)$. And there is a canonical such map:

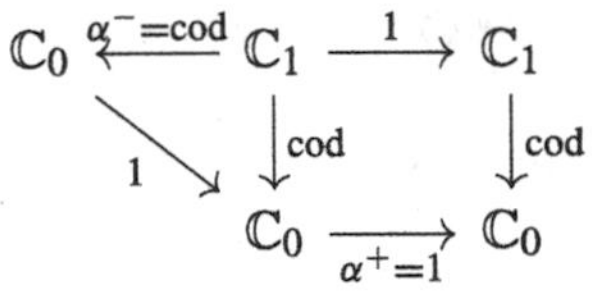

$$
\begin{array}{ccccc}
\mathbb{C}_0 & \xleftarrow{\alpha^-=\mathrm{cod}} & \mathbb{C}_1 & \xrightarrow{1} & \mathbb{C}_1 \\
& \searrow_{1} & \downarrow{\scriptstyle \mathrm{cod}} & & \downarrow{\scriptstyle \mathrm{cod}} \\
& & \mathbb{C}_0 & \xrightarrow[\alpha^+=1]{} & \mathbb{C}_0
\end{array}
$$

One readily checks this is indeed a strength and that with respect to this strength the induced comonad is strong.

9.2 A Simplicial Poset of Traversals

Let us define an internal poset $\mathbb{T}$ in simplicial sets.

The object of objects $\mathbb{T}_0$ has as its n-simplices the *n-dimensional traversals*. An n-dimensional traversal is a finite sequence of elements from $[n] \times \{+, -\}$, that is, a function $\theta : \{1, \ldots, l\} \to [n] \times \{+, -\}$ for some $l \in \mathbb{N}$ (including the empty traversal for $l = 0$). A good way to picture a traversal is as follows. An n-dimensional traversal is like a zigzag:

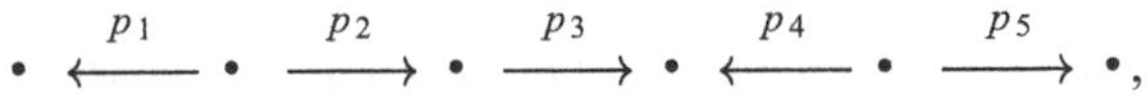

$$
\bullet \xleftarrow{p_1} \bullet \xrightarrow{p_2} \bullet \xrightarrow{p_3} \bullet \xleftarrow{p_4} \bullet \xrightarrow{p_5} \bullet,
$$

a (possibly empty) sequence of edges pointing either to the left (-) or right (+), with a label $p_i \in [n]$. The collection of such traversals is a simplicial set: the face map d_i acts on such a traversal by removing all the edges labelled with i and relabelling the other edges (meaning: if an edge is labelled with $j > i$, replace that label by $j-1$). The degeneracy s_i acts on such a traversal by duplicating edges labelled with i (with the copies pointing in the same direction as the original edge) and labelling the first copy $i+1$ and the second i in case the edge points to the right, and labelling the first copy i and the second $i+1$ if the edge point to the left. Other edges are relabelled accordingly (meaning: if an edge was labelled $j > i$, then it now has the label $j+1$). In general, the action by some $\alpha : [m] \to [n]$ on such a traversal θ is given as follows: if the label of some edge is i, then replace it by $\#\alpha^{-1}(i)$ many edges pointing in the same direction as the original edge, labelled by the elements of $\alpha^{-1}(i)$ in decreasing order if the edge points to the right and in increasing order if the edge points to the left. In short, $\theta \cdot \alpha$ is the unique map fitting into a pullback square

$$\begin{array}{ccc} \{1,\ldots,l\} & \xrightarrow{\theta\cdot\alpha} & [m]\times\{+,-\} \\ \downarrow{\scriptstyle v} & & \downarrow{\scriptstyle \alpha\times 1} \\ \{1,\ldots,k\} & \xrightarrow{\theta} & [n]\times\{+,-\} \end{array}$$

with $\mathrm{proj}_{[m]}.(\theta \cdot \alpha) : \{1,\ldots,l\} \to [m]$ decreasing on those fibres $v^{-1}(i)$ with $\theta(i)$ positive, and increasing on those fibres $v^{-1}(i)$ with $\theta(i)$ negative.

A *position* in an n-dimensional traversal $\theta : \{1,\ldots,l\} \to [n]\times\{+,-\}$ is a choice of one of the vertices: formally, it is an element $p \in \{0, 1, \ldots, l\}$. The elements of $(\mathbb{T}_1)_n$ are pairs consisting of an n-dimensional traversal θ together with a position in this traversal (a *pointed traversal*). The action of α on the traversals is as before, while it acts on the choice of vertex as follows: if $\theta' = \theta \cdot \alpha$, and v is some vertex in θ, then we choose that vertex in θ' which is the rightmost vertex in θ' which is either the source or target of an edge which is a copy of an edge which was to the left of v (choosing the leftmost vertex if no such edge exists).

There are two maps cod, dom : $\mathbb{T}_1 \to \mathbb{T}_0$ with cod being the obvious forgetful map (forgetting the choice of position), while dom removes the part of the traversal *before* the position. That means that we think of $\mathbb{T}$ as a simplicial poset with the final segment ordering ($\theta_0 \leq \theta_1$ if θ_0 is a final segment of θ_1: in that case, there is a position p in θ_1 such that after that point we see θ_0, and p can be thought of as the morphism from θ_0 to θ_1).

To see that this is an internal poset, note that there is a map id : $\mathbb{T}_0 \to \mathbb{T}_1$ given by choosing the position at the start of the traversal. Finally, we need a map

$$\mathrm{comp} : \mathbb{T}_1 \times_{\mathbb{T}_0} \mathbb{T}_1 \to \mathbb{T}_1.$$

That is, we start with pointed traversals (θ_1, p_1) and (θ_0, p_0) such that θ_0 is the final segment we obtain from θ_1 by removing everything before position p_1. Then comp takes θ_1 with position p_0 (which is a position in θ_0, and, because θ_0 is a final segment of θ_1, in θ_1 as well).

In view of the correspondence between internal categories and polynomial comonads, this internal category induces a polynomial comonad, whose counit we denote $s : \mathrm{cod} \to I$ and whose comultiplication we call $\Gamma : \mathrm{cod} \to \mathrm{cod} \otimes \mathrm{cod}$. Therefore the *simplicial Moore functor* $M = P_{\mathrm{cod}}$ defined as

$$(MX)_n = \sum_{\theta \in (\mathbb{T}_0)_n} X^{\mathrm{cod}^{-1}(\theta)}$$

carries the structure of a strong comonad with counit $s : M \Rightarrow 1$ and comultiplication $\Gamma : M \Rightarrow MM$.

Remark 9.1 Note that $M1 \cong \mathbb{T}_0$. In fact, the object $\mathbb{T}_0$ was introduced as $M1$ in [19].

9.3 Simplicial Moore Paths

M has more structure: in fact, we have an internal category in $\mathrm{Poly}(\mathcal{E})$. To see this, note that we can also equip $\mathbb{T}_0$ with the initial segment ordering. In that case, we take the same codomain map, but now as domain map we take $\mathrm{dom}^* : \mathbb{T}_1 \to \mathbb{T}_0$ which removes the part of the traversal *after* the chosen position. In addition, we have a map $\mathrm{id}^* : \mathbb{T}_0 \to \mathbb{T}_1$ which chooses the endpoint of the given traversal as its chosen position, as well as an appropriate composition

$$\mathrm{comp}^* : \mathbb{T}_1 \times_{\mathbb{T}_0} \mathbb{T}_1 \to \mathbb{T}_1.$$

This means that M carries a second strong comonad structure with counit $t : M \Rightarrow 1$ and comultiplication $\Gamma^* : M \Rightarrow MM$.

Note that with either ordering, the poset $\mathbb{T}$ has an initial object $0 : 1 \to \mathbb{T}_0$ (the unique traversal of length 0), and with the final segment ordering, the map $\mathrm{id}^* : \mathbb{T}_0 \to \mathbb{T}_1$ points to the unique map from the initial traversal to the given traversal (and similarly for id and the initial segment ordering). This means that we also have a map $r : I \to \mathrm{cod}$ given by:

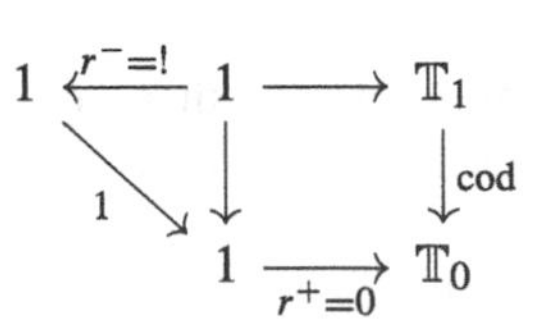

$$\begin{array}{ccccc} 1 & \xleftarrow{r^- = !} & 1 & \longrightarrow & \mathbb{T}_1 \\ & {\scriptstyle 1}\searrow & \downarrow & & \downarrow {\scriptstyle \mathrm{cod}} \\ & & 1 & \xrightarrow[r^+ = 0]{} & \mathbb{T}_0 \end{array}$$

Note that because r^- iso, the natural transformation induced by r is cartesian.

At this point one readily checks all the axioms for a two-sided Moore structure which do not involve the multiplication μ. In fact, all of these follow simply from the fact that we are working in an internal category with an initial object 0 with the property that the only map $C \to 0$ is the identity on 0.

1. The equation $s.r = t.r = 1$ follows immediately from the fact that there is only polynomial map $I \to I$.
2. r is strong: $\alpha.(1 \times r) = r.p_1 : X \times 1 \to MX$. In this we have to compare two maps $(1 \to 1) \times (0 \to 1) \to (\mathrm{cod} : \mathbb{T}_1 \to \mathbb{T}_0)$. In the forwards direction they are both $0 : 1 \to \mathbb{T}_0$, while in the backwards direction they are both equal as well, because they both have codomain 1.
3. $\Gamma.r = rM.r$, or: $\Gamma.r = (r \otimes 1_{\mathrm{cod}}).r$. We have $(\Gamma.r)^+ = \Gamma^+.r^+ = (0, \lambda\alpha : C \to 0.C)$, while $(r \otimes 1_{\mathrm{cod}}.r)^+ = (0, \lambda : C \to 0.0)$, which coincide, while in the backwards direction we again have to compare two maps which terminate in 1: so these are again equal.
4. $tM.\Gamma = r.t$, or $(t \otimes 1_{\mathrm{cod}}).\Gamma = r.t$. Note $(t \otimes 1_{\mathrm{cod}}.\Gamma)^+(C) = \mathrm{dom}(0 \to C) = 0$, while $(r.t)^+(C) = 0$. Also, $(r.t)^-_C(\alpha : D \to 0) = ! : 0 \to C$ and $((t \otimes 1_{\mathrm{cod}}).\Gamma)^-_C(\alpha : D \to 0) = \mathrm{comp}(! : 0 \to C, \alpha : D \to 0)$.
5. $Mt.\Gamma = \alpha.(t, M!) : \mathrm{cod} \to \mathrm{cod}$. Here $(t, M!) : \mathrm{cod} \to (1 : \mathbb{T}_0 \to \mathbb{T}_0)$ is given by $(1_{\mathbb{T}_0}, \mathrm{id}^*)$, so that the right hand side is $(1, \lambda\alpha : D \to C.! : 0 \to C)$. The left hand side, however, is given by $(Mt.\Gamma)^+(C) = (1_{\mathrm{cod}} \otimes t)(C, \lambda\alpha : D \to C.D) = C$, while $(Mt.\Gamma)^-_C(\alpha : D \to C) = \Gamma^-.(1_{\mathrm{cod}} \otimes t)^-(\alpha : D \to C) = \mathrm{comp}(\alpha : D \to C, ! : 0 \to D) = ! : 0 \to C$, as desired.
6. Equations similar to those in (3–5) have to be verified for Γ^* as well: but since also with the initial segment ordering, $\mathbb{T}$ has a strong initial object, the same arguments will work.

What is still needed, then, is to define $\mu_X : MX^t \times^s_X MX \to MX$ and to verify that it satisfies all the expected equations.

Using the formula for computing pullbacks of polynomials, we see that in order to define μ we need maps

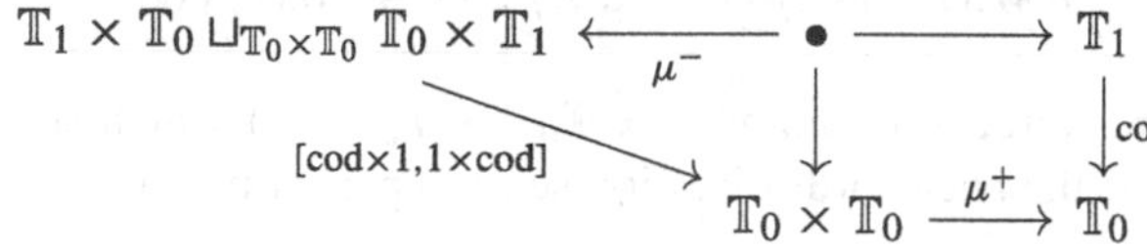

$$\begin{array}{ccccc}
\mathbb{T}_1 \times \mathbb{T}_0 \sqcup_{\mathbb{T}_0 \times \mathbb{T}_0} \mathbb{T}_0 \times \mathbb{T}_1 & \xleftarrow{\ \mu^-\ } & \bullet & \longrightarrow & \mathbb{T}_1 \\
 & \searrow^{[\mathrm{cod} \times 1, 1 \times \mathrm{cod}]} & \downarrow & & \downarrow \mathrm{cod} \\
 & & \mathbb{T}_0 \times \mathbb{T}_0 & \xrightarrow{\ \mu^+\ } & \mathbb{T}_0
\end{array}$$

Note that the fibre over (θ_0, θ_1) of the map on the left is the collection of positions in θ_0 and θ_1, with the final position in θ_0 identified with the initial position in θ_1. So what we can do is define $\mu^+(\theta_0, \theta_1) = \theta_0 * \theta_1$, the concatenation of the two sequences with θ_0 put before θ_1. Since the positions in $\theta_0 * \theta_1$ are precisely the positions in either θ_0 or θ_1, with the final position in θ_0 coinciding with the initial position in θ_1, we have a pullback square

$$\begin{array}{ccc} \mathbb{T}_1 \times \mathbb{T}_0 \sqcup_{\mathbb{T}_0 \times \mathbb{T}_0} \mathbb{T}_0 \times \mathbb{T}_1 & \xrightarrow{\mu^*} & \mathbb{T}_1 \\ {\scriptstyle [\mathrm{cod}\times 1, 1\times \mathrm{cod}]}\downarrow & & \downarrow{\scriptstyle \mathrm{cod}} \\ \mathbb{T}_0 \times \mathbb{T}_0 & \xrightarrow{\mu^+} & \mathbb{T}_0. \end{array}$$

So we can choose μ^- to be an isomorphism and μ will be a cartesian natural transformation.

We will leave it to the reader to verify that μ is strong and combines with r, s, t to yield a category structure, which is both left and right cancellative. The most difficult axioms to check are the distributive laws and the sandwich equation (see Definitions A.1 and A.2, respectively), which we will discuss here in some detail, also because they were not part of [19].

Lemma 9.1 *The distributive law*

$$\Gamma.\mu = \mu.(M\mu.\nu_X.(\Gamma.p_1, \alpha_{MX}.(p_2, M!.p_1)), \Gamma.p_2) : MX \times_X MX \to MMX$$

holds, as does the corresponding law for Γ^*.

Proof We only show the distributive law for Γ as the corresponding statement for Γ^* is proved similarly.

We have to compare two maps

$$\mathrm{cod} \times_I \mathrm{cod} \to \mathrm{cod} \otimes \mathrm{cod}.$$

The left hand side goes via cod and in the positive direction sends (θ_0, θ_1) to $(\theta_0 * \theta_1, \lambda p \in \mathrm{cod}^{-1}(\theta_0 * \theta_1).\mathrm{dom}(p))$, and in the negative direction sends a pair of positions p_0 in $\theta_0 * \theta_1$ and p_1 in $\mathrm{dom}(p_0)$ to the position corresponding to p_1 in either θ_0 or θ_1.

Let us now try to decompose the right hand side. The map

$$\alpha M.(p_2, M!.p_1) : \mathrm{cod} \times_I \mathrm{cod} \to \mathrm{cod} \otimes \mathrm{cod}$$

is in the forwards direction a map $\mathbb{T}_0 \times \mathbb{T}_0 \to P_{\mathrm{cod}}(\mathbb{T}_0)$ which sends (θ_0, θ_1) to $(\theta_0, \lambda p.\theta_1)$, while the backwards direction sends a pair of positions $(p_0 \in \theta_0, p_1 \in \theta_1)$ to p_1.

Then the map $M\mu.\nu.(\Gamma.p_1, \alpha.(p_2, M!.p_1))$ can be seen as a composition:

$$\begin{array}{ccc} \mathrm{cod} \times_I \mathrm{cod} \longrightarrow & (\mathrm{cod} \otimes \mathrm{cod}) \times_{\mathrm{cod}\otimes I} (\mathrm{cod} \otimes \mathrm{cod}) & \\ & \downarrow \cong & \\ & \mathrm{cod} \otimes (\mathrm{cod} \times_I \mathrm{cod}) & \xrightarrow{1_{\mathrm{cod}} \otimes \mu} \mathrm{cod} \otimes \mathrm{cod} \end{array}$$

where in the forwards directions these maps send (θ_0, θ_1) first to

$$((\theta_0, \lambda p \in \mathrm{cod}^{-1}(\theta_0).\mathrm{dom}(p)), (\theta_0, \lambda p \in \mathrm{cod}^{-1}(\theta_0).\theta_1),$$

which gets rewritten to $(\theta_0, \lambda p \in \mathrm{cod}^{-1}(\theta_0).(\mathrm{dom}(p), \theta_1)$, and then sent to

$$(\theta_0, \lambda p \in \mathrm{cod}^{-1}(\theta_0).\mathrm{dom}(p) * \theta_1).$$

In the backwards direction it sends a pair of positions in p_0 in θ_0 and p_1 in $\mathrm{dom}(p_0) * \theta_1$ first to the pair p_0 and the position corresponding to p_1 either in $\mathrm{dom}(p_0)$ or θ_1, and then to the position corresponding to p_1 in θ_0 if it belongs to $\mathrm{dom}(p_0)$ or to the position corresponding to p_1 in θ_1 if it belongs to θ_1. In short, it sends p_0 and p_1 to the position corresponding to p_1 in either θ_0 or θ_1.

In the final step we look at the whole right hand side as a composition

$$\begin{array}{ccc} \mathbf{cod} \times_1 \mathbf{cod} \longrightarrow (\mathbf{cod} \otimes \mathbf{cod}) \times_{1 \otimes \mathbf{cod}} (\mathbf{cod} \otimes \mathbf{cod}) & & \\ \downarrow \cong & & \\ (\mathbf{cod} \times_1 \mathbf{cod}) \otimes \mathbf{cod} & \longrightarrow & \mathbf{cod} \otimes \mathbf{cod}. \end{array}$$

In the positive direction this takes (θ_0, θ_1) first to

$$((\theta_0, \lambda p \in \mathrm{cod}^{-1}(\theta_0).\mathrm{dom}(p) * \theta_1), (\theta_1, \lambda p \in \mathrm{cod}^{-1}(\theta_1).\mathrm{dom}(p))),$$

and then to $(\theta_0 * \theta_1, \lambda p \in \mathrm{cod}^{-1}(\theta_0 * \theta_1).\mathrm{dom}(p))$, as before. In the backwards direction we are given a pair consisting of a position p_0 in $\theta_0 * \theta_1$ and a position p_1 in $\mathrm{dom}(p_0)$ and we start by making a case distinction on whether p_0 lies in θ_0 or θ_1. If it lies in θ_1, the pair gets mapped to the position p_1 in θ_1. If it lies in θ_0, the pair gets mapped to the position corresponding to p_1 in either θ_0 or θ_1. So in either case it gets mapped to the position corresponding to p_1 in either θ_0 or θ_1, as before. □

Lemma 9.2 *The sandwich equation* $M\mu.\nu.(\Gamma^*, \Gamma) = \alpha M.(1, M!) : M \to MM$ *holds.*

Proof We have to compare two morphisms $\mathbf{cod} \to \mathbf{cod} \otimes \mathbf{cod}$. The right hand side can be seen as a composition:

$$\mathbf{cod} \longrightarrow \mathbf{cod} \times 1_{\mathbb{T}_0} \xrightarrow{\cong} 1_{\mathbb{T}_0} \otimes \mathbf{cod} \xrightarrow{\alpha \otimes 1_{\mathbf{cod}}} \mathbf{cod} \otimes \mathbf{cod}.$$

In the positive direction these maps are the diagonal $\mathbb{T}_0 \to \mathbb{T}_0 \times \mathbb{T}_0$, a map $\mathbb{T}_0 \times \mathbb{T}_0 \to \mathbb{T}_0 \times \mathbb{T}_0$ swapping the two arguments and a map $\mathbb{T}_0 \times \mathbb{T}_0 \to P_{\mathbf{cod}}(\mathbb{T}_0)$ sending (θ_0, θ_1) to $(\theta_0, \lambda p.\theta_1)$. In short, in the positive direction this maps sends θ to $(\theta, \lambda p.\theta)$. In the negative direction, a pair of positions (p, p') in θ is sent to p'.

The left hand side can be seen as a composition

$$\begin{array}{ccc} \text{cod} \longrightarrow (\text{cod} \otimes \text{cod}) \times_{\text{cod}} (\text{cod} \otimes \text{cod}) & & \\ \downarrow \cong & & \\ \text{cod} \otimes (\text{cod} \times_I \text{cod}) & \xrightarrow{1_{\text{cod}} \otimes \mu} & \text{cod} \otimes \text{cod}. \end{array}$$

In the positive direction this first sends θ to $(\theta, \text{dom}^*, \text{dom})$ and then it sends (θ, t, t') to $(\theta, \lambda p.t(p) * t'(p))$. So the composition is $(\theta, \lambda p.\theta)$: the reason is that dom* removes the part after the position, dom removes the part before the position, and * concatenates the results: so we just get the original traversal back. In the backwards direction a pair of positions (p, p') is first sent to the position corresponding to p' in either the part before or after p, and then it is sent to the corresponding position in the whole traversal. In short, it is sent to p'. □

This finishes the verification of the axioms for a two-sided Moore structure. Note that all the proofs that we have given so far would still work if we restricted the traversals in $\mathbb{T}_0$ to those which only move towards the right (that is, those traversals $\theta : \{1, \ldots, l\} \to [n] \times \{+, -\}$ for which $\theta(i)$ for any $i \in \{1, \ldots, l\}$ is always of the form $(k, +)$ for some $k \in [n]$). We will refer to the version of M that we get in this way as M_+. Of course, similar remarks apply if we restrict the traversals to those that only move to the left; the version of M that we would have obtained in that way will be referred to as M_-.

Theorem 9.1 *The endofunctors M, M_+ and M_- equip the category of simplicial sets with three distinct two-sided Moore structures.*

It remains to check that M equips the category of simplicial sets with the structure of a symmetric Moore category. This means that we should be able to construct a twist map τ if we work with two different orientations. Indeed, in that case there is a map of polynomials

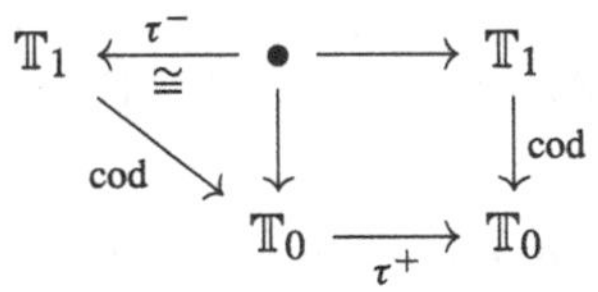

$$\begin{array}{ccccc} \mathbb{T}_1 & \xleftarrow[\cong]{\tau^-} & \bullet & \longrightarrow & \mathbb{T}_1 \\ & \searrow^{\text{cod}} & \downarrow & & \downarrow \text{cod} \\ & & \mathbb{T}_0 & \xrightarrow[\tau^+]{} & \mathbb{T}_0 \end{array}$$

with τ^+ sending a traversal $\theta : \{1, \ldots, l\} \to [n] \times \{+, -\}$ to a traversal $\tau^+(\theta)$ with the same length l and $\tau^+(\theta)(i) = \tau(\theta(l + 1 - i))$, where $\tau(k, +) = (k, -)$ and $\tau(k, -) = (k, +)$ for any $k \in [n]$. So τ^+ reverses the order and orientation of the traversal. Finally, τ^- sends a position p in such a traversal to the position $l - p$. Note that τ^- is an isomorphism and τ is a cartesian natural transformation.

Most of the equations for τ are easy to verify, except perhaps for $\Gamma^* = \tau M.M\tau.\Gamma.\tau$, which is equivalent to $\Gamma^*.\tau = \tau M.M\tau.\Gamma$, or $\Gamma^*.\tau = (\tau \otimes \tau).\Gamma$. In the positive direction we have that $\Gamma^*.\tau$ is the function sending a traversal θ to the pair where the first component is this traversal reversed, while the second component is the function which takes a position in this reversed traversal and removes the part in this reversed traversal after this position. In the backwards direction this takes a position in the reversed traversal and a position in the reversed traversal before the first position and produces the position corresponding to the second position in the original traversal. The map $\tau \otimes \tau$ is the function which in the positive direction takes a pair $(\theta, t : \text{cod}^{-1} \to \mathbb{T}_0)$ to $(\tau^+(\theta), \tau^+ \circ t \circ \tau^-)$, so if t is the domain function (removing the part of the traversal before the given position), this agrees with the other function in the positive direction. In the backwards direction, $(\tau \otimes \tau).\Gamma$ takes a position in the reversed traversal and a position in the reversed traversal before the given position, first reverses both and then takes the position corresponding to the second position in the original traversal, as before.

This means that we have the following result, as promised:

Theorem 9.2 *The endofunctor M equips the category of simplicial sets with the structure of a symmetric Moore category.*

In view of the results of Sect. 4.2, Theorem 9.1 implies that there are *a priori* six AWFSs on the category of simplicial sets. However, up to isomorphism, there are only three, because

- the twist map τ induces an isomorphism between the AWFS determined by (M_+, Γ_+, s) and the one determined by (M_-, Γ^*_-, t).
- the twist map τ induces an isomorphism between the AWFS determined by (M_+, Γ^*_+, t) and the one determined by (M_-, Γ_-, s).
- the Moore structure determined by M is symmetric, so the AWFS determined by (M, Γ, s) is isomorphic to the one determined by (M, Γ^*, t).

In view of this we will make the following definition:

Definition 9.2 We will refer to the algebras for the monads of the three AWFSs above as *naive right fibrations*, *naive left fibrations* and *naive Kan fibrations*, respectively. We will refer to the coalgebras for the comonad of the third AWFS as *HDRs*.

9.4 Geometric Realization of a Traversal

In this chapter we have defined M as the polynomial functor associated to cod : $\mathbb{T}_1 \to \mathbb{T}_0$. In [19], the Moore path functor was defined differently (as a parametric right adjoint). The main goal of this section is to show that the two descriptions are equivalent. Our proof here will be fairly combinatorial; a more conceptual proof can be found in the second author's PhD thesis [44] (see Remark 9.2 below).

Our combinatorial argument makes us of the "geometric realization" of a traversal, a construction which can already be found in [19], and will also play an important role in the later chapters.

Definition 9.3 For an element of the form $(k, \pm) \in [n] \times \{+, -\}$, let us define: $(k,+)^s = k+1$, $(k,+)^t = k$, $(k,-)^s = k$, $(k,-)^t = k+1$. If θ is a n-dimensional traversal of length k, then we define its *geometric realization* $\widehat{\theta}$ to be the colimit of the diagram

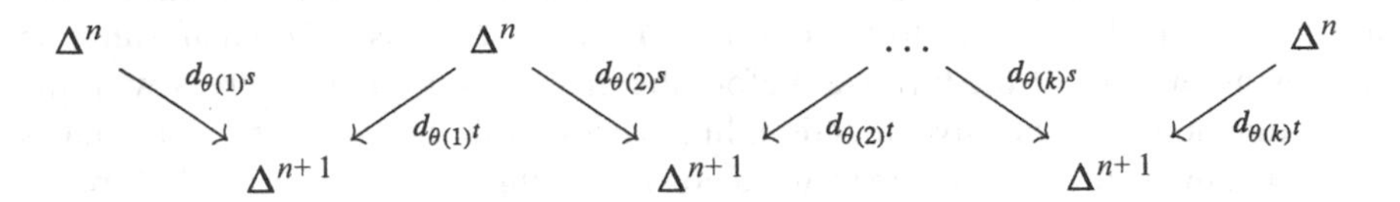

in simplicial sets. In words: we turn an n-dimensional traversal into a simplicial set, by replacing its vertices by n-simplices and its edges by $(n+1)$-simplices, in such a way that if an $(n+1)$-simplex comes from the ith edge, then the n-simplices coming from the vertices connected by that edge are its $\theta(i)^s$-th and $\theta(i)^t$-th faces, respectively.

Theorem 9.3 *The geometric realization $\widehat{\theta}$ of an n-dimensional traversal θ fits into a pullback square*

$$
\begin{array}{ccc}
\widehat{\theta} & \xrightarrow{k_\theta} & \mathbb{T}_1 \\
\downarrow{\scriptstyle j_\theta} & & \downarrow{\scriptstyle \mathrm{cod}} \\
\Delta^n & \xrightarrow{\theta} & \mathbb{T}_0.
\end{array}
\tag{9.1}
$$

Proof We have to construct two maps $j_\theta : \widehat{\theta} \to \Delta^n$ and $k_\theta : \widehat{\theta} \to \mathbb{T}_1$, which we will do using that $\widehat{\theta}$ is a colimit. If $\theta(i) = (k, \pm)$, let us write $\overline{\theta(i)} = k$. Then

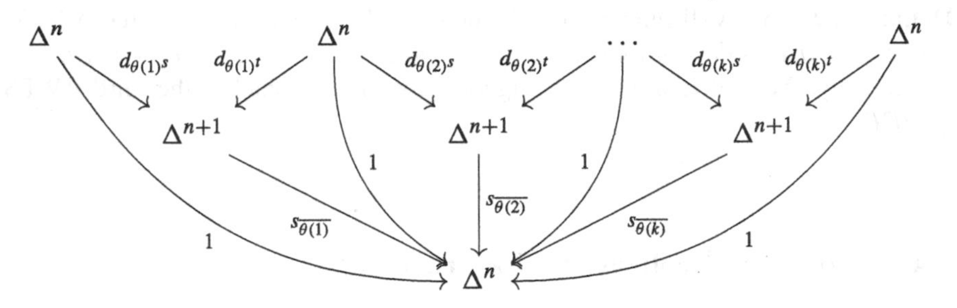

commutes, so determines a map $j_\theta : \widehat{\theta} \to \Delta^n$. In addition, we can construct cocone with vertex $\mathbb{T}_1$:

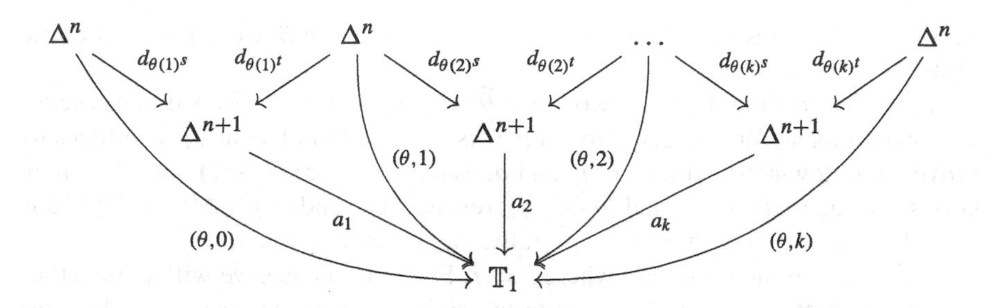

In this diagram the n-simplices along the top correspond to positions p in θ, which correspond to maps $(\theta, p) : \Delta^n \to \mathbb{T}_1$. Furthermore, the ith $(n+1)$-simplex in the second row comes from the ith edge in θ and for this $(n+1)$-simplex we choose a position in $\theta \cdot s_{\overline{\theta(i)}}$: note that that in $\theta \cdot s_{\overline{\theta(i)}}$ the edge in question gets duplicated and the position we choose is the one inbetween the two copies (we will refer to this as a "special position"). This determines maps $a_i : \Delta^{n+1} \to \mathbb{T}_1$ which make the diagram above commute, and hence we obtain a map $k_\theta : \widehat{\theta} \to \mathbb{T}_1$. Again using that $\widehat{\theta}$ is a colimit, it is not hard to see that with the resulting maps the square in the statement of the theorem commutes. It remains to show that it is a pullback.

Imagine that we start with a map $\alpha : \Delta^m \to \Delta^n$ and a position p in $\theta \cdot \alpha$. Our first task is to show there is some m-simplex in $\widehat{\theta}$ which gets mapped to α and p by the maps we have just constructed. Let us partition the edges in $\theta \cdot \alpha$ by grouping together those edges which come from the same edge in θ: we will call these groups "blocks". In other words, they are the fibres of the map v as in the pullback square

$$\begin{array}{ccc} \{1, \ldots, l\} & \xrightarrow{\theta \cdot \alpha} & [m] \times \{+, -\} \\ \downarrow{\scriptstyle v} & & \downarrow{\scriptstyle \alpha \times 1} \\ \{1, \ldots, k\} & \xrightarrow{\theta} & [n] \times \{+, -\} \end{array}$$

determining $\theta \cdot \alpha$. For the position p there are now two possibilities: the first is that p is the boundary between two blocks (that is, the edges to the right and left of p come from different edges in θ; this is meant to include the case where p is one of the outer positions). In that case there is some position q in θ such that p is the restriction of q along α. Then we have the map $(\theta, q) : \Delta^n \to \mathbb{T}_1$ corresponding to the position q, which we can also regard as an n-simplex in $\widehat{\theta}$ lying over the identity in Δ^n. Restricting this one along α, we get an element in $\widehat{\theta}$ of the form we want. The other case is that p belongs to the interior of one of the blocks with the edges to the left and right mapping to the same edge with label i in θ. In that case we can use that if $\alpha : [m] \to [n]$ is some map in $\mathbf{\Delta}$, and $\alpha(k) = \alpha(k+1) = i$, then there is a map $\beta : [m] \to [n+1]$ such that $\alpha = s_i.\beta$ and $\beta(k) \neq \beta(k+1)$. (Indeed, we can put $\beta(j) = \alpha(j)$ if $j \leq k$ and $\beta(j) = \alpha(j) + 1$ if $j > k$.) Then there is some special position q in $\theta \cdot s_i$ such that p is the restriction of q along β. This determines a map $a_j : \Delta^{n+1} \to \mathbb{T}_1$, which we can also regard as an $(n+1)$-simplex in $\widehat{\theta}$ lying over s_i in Δ^n. By restricting this $(n+1)$-simplex in $\widehat{\theta}$ along β, we get an m-simplex of the form we want.

It remains to show that the two maps $\widehat{\theta} \to \Delta^n$ and $\widehat{\theta} \to \mathbb{T}_1$ we constructed are jointly monic. Since every element in $\widehat{\theta}$ is a restriction of some a_i, it suffices to prove the following statement: if a_u and a_v with $u \leq v$ are $(n+1)$-simplices in $\widehat{\theta}$ corresponding to $(\theta \cdot s_i, p)$ and $(\theta \cdot s_j, q)$, respectively, and $\alpha, \beta : \Delta^m \to \Delta^{n+1}$ are such that $s_i.\alpha = s_j.\beta$ and $p \cdot \alpha = q \cdot \beta$, then $a_u \cdot \alpha = a_v \cdot \beta$ in $\widehat{\theta}$.

Let us first consider the case where $i \neq j$. For convenience, we will assume that $i < j$, and both edges i and j point to the right. Then we can take the following pullback:

$$\begin{array}{ccc} \Delta^{n+2} & \xrightarrow{s_i} & \Delta^{n+1} \\ \downarrow{\scriptstyle s_{j+1}} & & \downarrow{\scriptstyle s_j} \\ \Delta^{n+1} & \xrightarrow{s_i} & \Delta^n. \end{array}$$

So there is some map γ such that $\alpha = s_{j+1}.\gamma$ and $\beta = s_i.\gamma$ and we obtain the equation $p \cdot s_{j+1}.\gamma = q \cdot s_i.\gamma$. Note that both $p \cdot s_{j+1}$ and $q \cdot s_i$ are distinct positions in the same traversal and what the equation is saying is that they become identified after restricting along γ. The crucial observation is that this can only happen if γ removes the edge to the right of p and the one to the left of q and everything else inbetween. In particular, γ factors through d_i and d_{j+2} and we can write $\gamma = d_{j+2}.d_i.\gamma' = d_i.d_{j+1}.\gamma'$, so that $\alpha = d_i.\gamma'$ and $\beta = d_{j+1}.\gamma'$. Since γ' must omit all the labels of edges between inbetween u and v, the following diagram commutes:

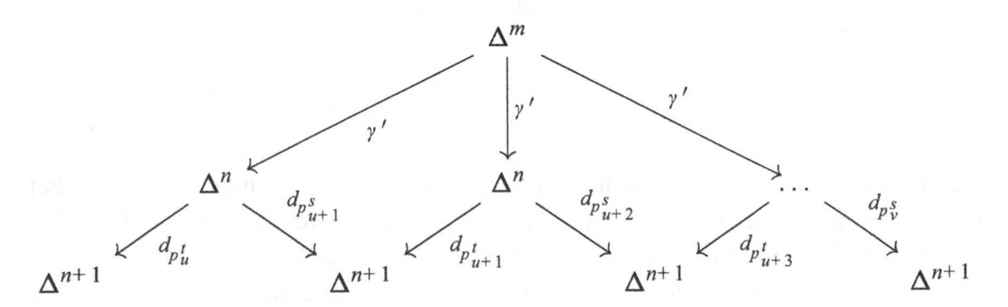

Since the composite along the left is α and composite along the right is β, this shows that $a_u \cdot \alpha = a_v \cdot \beta$ in $\widehat{\theta}$, as desired. (There are other cases to be considered: different directions and $i > j$, but it all works out.)

Let us now consider the case where $i = j$, but $u < v$. From $s_i.\alpha = s_i.\beta$, it follows that $\alpha^{-1}\{i, i+1\} = \beta^{-1}\{i, i+1\} = [k, l]$ for some k and l, whilst on inputs outside the interval $[k, l]$ the functions α and β are identical. In addition, we have the equation $p \cdot \alpha = q \cdot \beta$, which implies that α must omit i and β must omit $i+1$ (if, for simplicity, we assume that both edges point to the right; other cases are again similar). The reason is that outside the i-blocks (that is, outside the pairs of consecutive edges in $\theta \cdot s_i$ of the form $v^{-1}(e)$ with $\theta(e) = (i, \pm)$) restricting along

α and β acts in the same way. But also on these i-blocks α and β act in very similar ways: both replace them by strings of edges of length $l-k$ with identical labels. The only difference is that they may disagree on how to shift the special position. Hence to make the two positions coincide one therefore has to shift the chosen position to the endpoints of the blocks and to eliminate the stuff inbetween. So $\alpha = d_i.\gamma$ and $\beta = d_{i+1}.\delta$ and hence $\gamma = \delta$. Since $\gamma = \delta$ must also omit every label inbetween the edges u and v, we can again show (as in the previous case) that $a_u \cdot \alpha = a_v \cdot \beta$ in $\widehat{\theta}$.

Finally, if $u = v$, then $s_i.\alpha = s_i.\beta$ and $p \cdot \alpha = p \cdot \beta$. The former equation again implies that $\alpha^{-1}\{i, i+1\} = \beta^{-1}\{i, i+1\} = [k, l]$ for some k and l, whilst on inputs outside the interval $[k, l]$ the functions α and β are identical. But then the second equation implies that also on the i-blocks α and β act in the same way and hence α and β also agree on the interval $[k, l]$. Hence $\alpha = \beta$, and $a_u \cdot \alpha = a_v \cdot \beta$. □

Corollary 9.1 *Geometric realization is part of a functor*

$$\widehat{(-)} : \int_\Delta \mathbb{T}_0 \to \widehat{\Delta}.$$

In fact, writing $U : \int_\Delta \mathbb{T}_0 \to \widehat{\Delta}$ *for the functor sending* (n, θ) *to* Δ_n*, we may regard the* j_θ *from the previous theorem as the components of a cartesian natural transformation* $j : \widehat{(-)} \to U$.

Proof If θ is an n-dimensional traversal and $\alpha : \Delta^m \to \Delta^n$ is some map in Δ, then we have two pullbacks

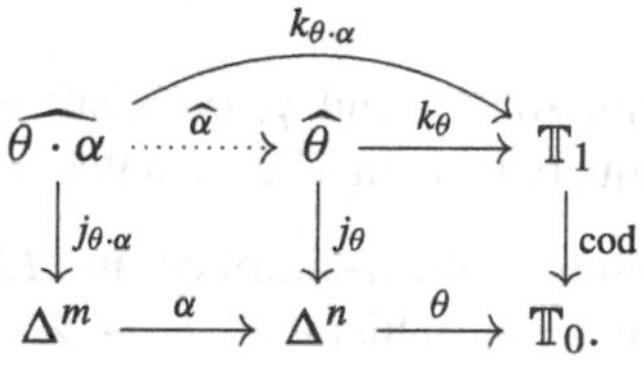

$$\begin{array}{ccccc}
 & & \overset{k_{\theta\cdot\alpha}}{\frown} & & \\
\widehat{\theta\cdot\alpha} & \overset{\widehat{\alpha}}{\dashrightarrow} & \widehat{\theta} & \xrightarrow{k_\theta} & \mathbb{T}_1 \\
\downarrow {\scriptstyle j_{\theta\cdot\alpha}} & & \downarrow {\scriptstyle j_\theta} & & \downarrow {\scriptstyle \mathrm{cod}} \\
\Delta^m & \xrightarrow{\alpha} & \Delta^n & \xrightarrow{\theta} & \mathbb{T}_0.
\end{array}$$

Therefore there exists a dotted arrow, turning the left hand square into a pullback as well. □

Corollary 9.2 *For the simplicial Moore path functor M we have*

$$(MX)_n \cong \sum_{\theta \in \mathbb{T}_0(n)} \mathrm{Hom}_{\widehat{\Delta}}(\widehat{\theta}, X).$$

Therefore the description given in this book is equivalent to the one given in [19].

Proof This is immediate from the following description of polynomial functors in presheaf categories (see [45], for instance): if $f : B \to A$ is a morphism of presheaves over $\mathbb{C}$, then

$$P_f(X)(C) = \sum_{a \in A(C)} \mathrm{Hom}_{\widehat{\mathbb{C}}}(B_a, X),$$

where B_a is the pullback

$$\begin{array}{ccc} B_a & \longrightarrow & B \\ \downarrow & & \downarrow f \\ yC & \xrightarrow[a]{} & A. \end{array}$$

□

Remark 9.2 As said, a more abstract proof of the previous theorem and two corollaries can be found in the second author's PhD thesis [44]. It relies on proving the categorical fact that when $P : C^{\mathrm{op}} \to Sets$ is a presheaf, $U : \mathcal{E} \to Sets$ is its category of elements, and $F : \mathcal{E} \to [C^{\mathrm{op}}, Sets]$ is a functor with a *cartesian* natural transformation $\gamma : F \Rightarrow y_{U \circ (-)}$, then the square

$$\begin{array}{ccc} F(e) & \longrightarrow & \mathrm{colim}\, F \\ \downarrow & & \downarrow \\ y_C & \longrightarrow & P \cong \mathrm{colim}\, y_{U \circ (-)} \end{array}$$

induced by the cocones (horizontal) and γ (vertical) is a pullback square. This square corresponds to the square (9.1) in Theorem 9.3.

For this reason we can think of the n-simplices in MX as pairs consisting of an n-dimensional traversal θ and a morphism $\pi : \widehat{\theta} \to X$ of presheaves. We think of these objects as *Moore paths* in X. These Moore path generalise ordinary paths in X, that is, maps $\mathbb{I} \to X$, in the following way.

We have the 0-dimensional traversals $\iota^+ =< (0, +) >$ and $\iota^- =< (0, -) >$ which correspond to two global sections

$$\iota^+, \iota^- : 1 \cong \Delta^0 \to \mathbb{T}_0.$$

Since $\mathbb{I} = \Delta^1$ is the geometric realization of both of these traversals, the previous theorem tells us that we have pullback squares

$$\begin{array}{ccc} \mathbb{I} & \longrightarrow & \mathbb{T}_1 \\ \downarrow & & \downarrow \mathrm{cod} \\ 1 & \xrightarrow[\iota^+/\iota^-]{} & \mathbb{T}_0. \end{array}$$

Regarding these pullback squares as morphisms of polynomials, we obtain two monic cartesian natural transformations $\iota^+, \iota^- : X^{\mathbb{I}} \to MX$. Furthermore, if we write

$$s, t : X^{\mathbb{I}} \to X$$

for the maps induced by $d_1 : \Delta^0 \to \Delta^1$ and $d_0 : \Delta^0 \to \Delta^1$, respectively, then we have that the following diagrams serially commute:

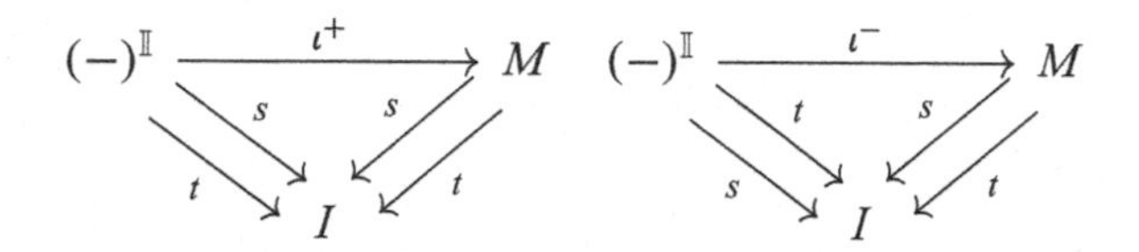

(So ι^+ preserves source and target, while ι^- reverses them.)

Remark 9.3 Another way of seeing that the usual path object $X^{\mathbb{I}}$ is a subobject of MX is as follows. We can take the pullback of the square above along a map from a representable $\Delta^n \to 1$:

$$\begin{array}{ccccc} \Delta^n \times \mathbb{I} & \longrightarrow & \mathbb{I} & \longrightarrow & \mathbb{T}_1 \\ \downarrow & & \downarrow & & \downarrow \scriptstyle\mathrm{cod} \\ \Delta^n & \longrightarrow & 1 & \xrightarrow[\iota^+/\iota^-]{} & \mathbb{T}_0. \end{array}$$

Indeed, what this says is that $\Delta^n \times \Delta^1$ is the geometric realisation of the traversals

$$< (n, +), (n-1, +), \ldots, (2, +), (1, +), (0, +) > \text{ and } < (0, -), (1, -), (2, -), \ldots, (n-1, -), (n, -) > .$$

Indeed, this reflects the well-known decomposition of the "prism" $\Delta^n \times \Delta^1$ as the union of $n+1$ many $(n+1)$-simplices; from our present point of view, this means that $\Delta^n \times \Delta^1$ occurs as the geometric realisation of these traversals. From this and the description of M in Corollary 9.2, one can also see that $X^{\mathbb{I}}$ embeds in MX.

Chapter 10
Hyperdeformation Retracts in Simplicial Sets

In the previous chapter we have shown that the endofunctor M equips the category of simplicial sets with symmetric Moore structure. Consequently, the category of simplicial sets carries an AWFS, with the coalgebras for the comonad being called the *HDRs* and the algebras for the monad being called the *naive Kan fibrations*. The purpose of this chapter is to take a closer look at this AWFS.

By definition, the naive Kan fibrations are generated by the large double category of HDRs. One important result in this chapter is that they are also generated by a small (countable) double category of HDRs, and that the naive fibrations Kan form a local notion of fibred structure. It should be apparent from the proofs that similar results would be true for naive left and right fibrations as well (see Definition 9.2).

10.1 Hyperdeformation Retracts Are Effective Cofibrations

Let us start by proving that HDRs are effective cofibrations. As we have seen in Lemma 6.3, for this it suffices to prove the following result:

Proposition 10.1 *The map $r_X : X \to MX$ is always an effective cofibration.*

Proof In fact, since

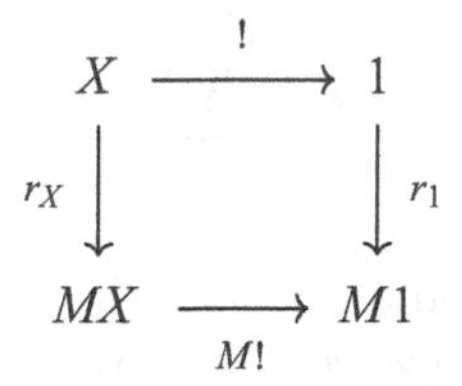

B. van den Berg, E. Faber, *Effective Kan Fibrations in Simplicial Sets*,
Lecture Notes in Mathematics 2321,
https://doi.org/10.1007/978-3-031-18900-5_10

is a pullback (r is cartesian), it suffices to prove this statement for $X = 1$. In other words, we have to define a map $\rho : \mathbb{T}_0 \cong M1 \to \Sigma$ such that

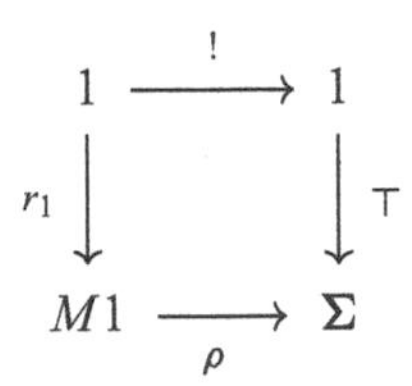

is a pullback. We set

$$\rho_n(\theta) = \{e \subseteq \{0, 1, \ldots, n\} \,:\, e \cap \mathrm{Im}(\theta) = \emptyset\}.$$

(So we take those subsets e for which no $i \in e$ occurs as $(i, +)$ or $(i, -)$ in the image of the traversal θ. This happens precisely when the restriction of the traversal θ along e is the unique traversal of length 0.) This is easily seen to be correct. □

10.2 Hyperdeformation Retracts as Internal Presheaves

In the sequel we will often have to prove that certain maps are HDRs. It turns out that for this purpose it is often convenient to use an equivalent description of the category of HDRs and morphisms of HDRs. Indeed, we have:

Theorem 10.1 *The category of HDRs in simplicial sets is equivalent to the category of internal presheaves on* $\mathbb{T}$.

Proof This is immediate from the fact that cod : HDR $\to$ $M-$Coalg is an equivalence (see Proposition 4.4) and the fact that M-coalgebras are equivalent to internal presheaves over $\mathbb{T}$. □

Let us unwind a bit more what this means and explain how one passes from HDRs to internal presheaves over $\mathbb{T}$ and back. If $(i : A \to B, j, H)$ is an HDR, then we can transpose

$$B \to MB = \sum_{\theta \in \mathbb{T}_0} B^{\mathrm{cod}^{-1}(\theta)}$$

to a pair of maps $d : B \to \mathbb{T}_0$ and $\rho : B\,{}^{d}\!\times_{\mathbb{T}_0}^{\mathrm{cod}} \mathbb{T}_1 \to B$ which satisfy the axioms for an internal presheaf over $\mathbb{T}$. Conversely, suppose one is given an internal presheaf over $\mathbb{T}_0$, that is, a pair of maps $d : B \to \mathbb{T}_0$ and $\rho : B \times_{\mathbb{T}_0} \mathbb{T}_1 \to B$ satisfying the right equations. The map $H : B \to MB$ is the transpose of ρ, while the inclusion $i : A \to B$ is the pullback

$$\begin{array}{ccc} A & \xrightarrow{i} & B \\ \downarrow{\scriptstyle !} & & \downarrow{\scriptstyle d} \\ 1 & \xrightarrow{0} & \mathbb{T}_0, \end{array}$$

so A is the fibre over the initial object in $\mathbb{T}$. Finally, $t.H : B \to B$ sends an element in B over θ to its restriction along the unique map $0 \to \theta$, and hence j does the same.

In the sequel we will often use this internal presheaf perspective on HDRs (see in particular the proofs of Propositions 10.2 and 10.3). The reason for this is that presheaves are easier to manipulate than HDRs, because they get rid of the exponential in the definition of MB and they (seemingly) contain less data. This becomes quite apparent when one considers the generic HDR from Proposition 10.2 which we will heavily exploit to construct more examples of HDRs. Therefore we will now translate the main constructions on HDRs (pullback, pushout, and vertical composition) into the language of internal presheaves.

Pullback Suppose $d : B \to \mathbb{T}_0$ and $\rho : B \times_{\mathbb{T}_0} \mathbb{T}_1 \to B$ is an internal presheaf and A is the fibre over 0 (that is, the pullback as above). If we are given a map $a : A' \to A$, then we obtain a new presheaf (B', d', ρ') by pullback as follows. First of all we take the pullback

$$\begin{array}{ccc} B' & \xrightarrow{b} & B \\ \downarrow{\scriptstyle j'} & & \downarrow{\scriptstyle j} \\ A' & \xrightarrow{a} & A. \end{array}$$

If $H : B \to MB$ is the map induced by d, ρ, then H' is the unique map filling

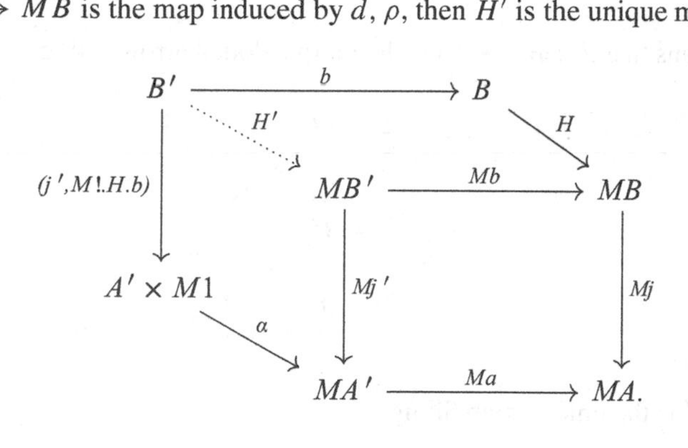

So $d' := M!.H' = M!.H.b = d.b : B' \to \mathbb{T}_0$. Moreover, ρ' will be the unique dotted arrow filling

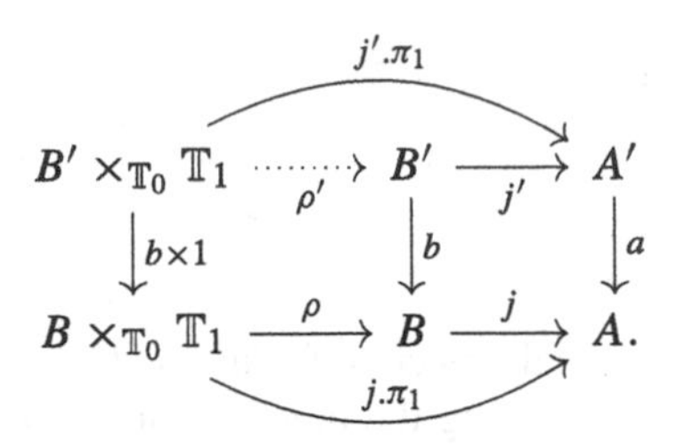

Pushout Again, suppose $d : B \to \mathbb{T}_0$ and $\rho : B \times_{\mathbb{T}_0} \mathbb{T}_1 \to B$ is an internal presheaf and A is the fibre over 0. If we are given a map $a : A \to A'$, we obtain a presheaf (B', d', ρ') by pushout, as follows. First, we take the pushout:

$$\begin{array}{ccc} A & \xrightarrow{a} & A' \\ \downarrow{\scriptstyle i} & & \downarrow{\scriptstyle i'} \\ B & \xrightarrow{b} & B'. \end{array}$$

If $H : B \to MB$ is the map induced by d, ρ, then $H' : B' \to MB'$ is the unique dotted arrow filling

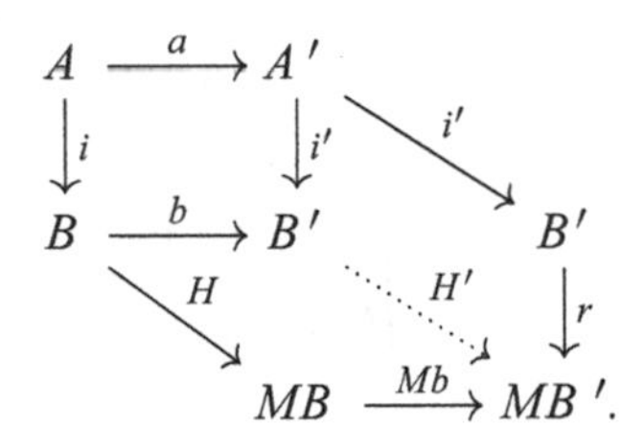

This means that $d' : B' \to \mathbb{T}_0$ is the unique dotted arrow filling

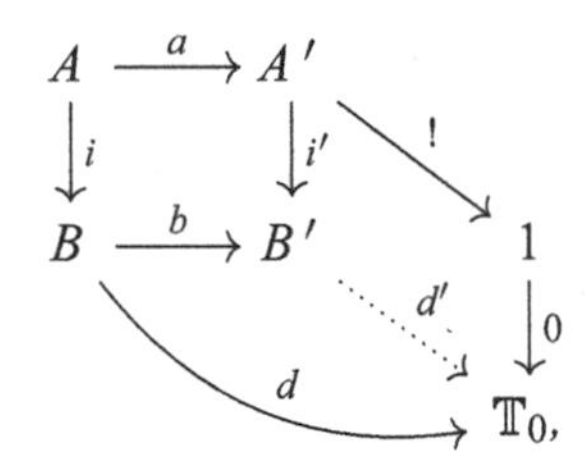

whilst ρ' is the unique map filling

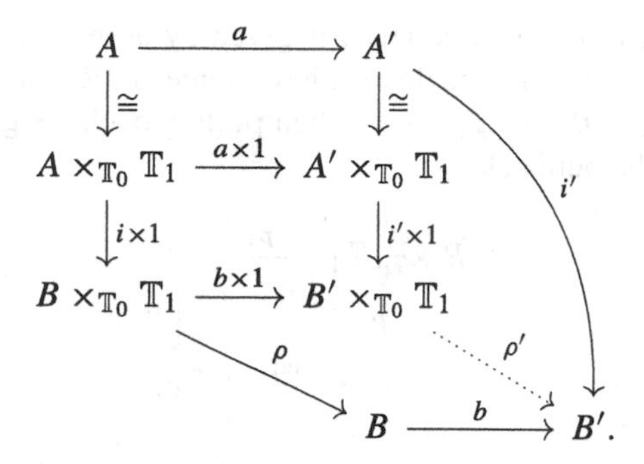

Vertical composition Suppose $(i_0 : A \to B, j_0, H_0)$ and $(i_1 : B \to C, j_1, H_1)$ are HDRs coming from presheaf structures (d_0, ρ_0) and (d_1, ρ_1). Then the vertical composition is given by

$$(i_1.i_0 : A \to C,\ j_0.j_1,\ \mu.(H_1, Mi_1.H_0.j_1)).$$

So this means we have a function $d_2 : C \to \mathbb{T}_0$, which is:

$$\begin{aligned} d_2 &= M!.\mu.(H_1, Mi_1.H_0.j_1) \\ &= \mu.(M!.H_1, M!.Mi_1.H_0.j_1) \\ &= \mu.(d_1, d_0.j_1) \\ &= d_1 * d_0.j_1. \end{aligned}$$

In addition, we need a morphism $\rho_2 : C \times_{\mathbb{T}_0} \mathbb{T}_1 \to C$. Here the domain can also be computed in two steps by taking the following two pullbacks:

$$\begin{array}{ccc} C \times_{\mathbb{T}_0} \mathbb{T}_1 \cup C \times_{\mathbb{T}_0} \mathbb{T}_1 & \xrightarrow{[\pi_1,\pi_1]} & C \\ \big\downarrow{\scriptstyle (\pi_2, d_0.j_1.\pi_1)\cup(d_1\times 1)} & & \big\downarrow{\scriptstyle (d_1, d_0.j_1)} \\ \mathbb{T}_1 \times \mathbb{T}_0 \cup \mathbb{T}_0 \times \mathbb{T}_1 & \xrightarrow{[\mathrm{cod}\times 1, 1\times \mathrm{cod}]} & \mathbb{T}_0 \times \mathbb{T}_0 \\ \big\downarrow{\scriptstyle \mu^*} & & \big\downarrow{\scriptstyle *} \\ \mathbb{T}_1 & \xrightarrow{\mathrm{cod}} & \mathbb{T}_0. \end{array}$$

Hence we can define ρ_2 as $[\rho_1, i_1.\rho_0.(j_1 \times 1)] : C \times_{\mathbb{T}_0} \mathbb{T}_1 \cup C \times_{\mathbb{T}_0} \mathbb{T}_1 \to C$.

We finish this section with the proof that the category of HDRs contains a "generic" element.

Proposition 10.2 *The triple* $(\mathrm{id}^* : \mathbb{T}_0 \to \mathbb{T}_1, \mathrm{cod}, \widehat{\mathrm{comp}})$ *is an HDR, which is generic in the following sense: for any HDR* $(i : A \to B, j, H)$ *there exists a pullback* (i', j', H') *of the generic one together with a morphism of HDRs* $(i', j', H') \to (i, j, H)$ *which is an epimorphism on the level of presheaves.*

Proof As a presheaf, the generic HDR is given by $\mathrm{dom} : \mathbb{T}_1 \to \mathbb{T}_0$ with $\mathrm{comp} : \mathbb{T}_1 \times_{\mathbb{T}_0} \mathbb{T}_1 \to \mathbb{T}_1$. Now imagine that we have some HDR, considered as a presheaf $d : B \to \mathbb{T}_0$ with $\rho : B \times_{\mathbb{T}_0} \mathbb{T}_1 \to B$. Then pulling back the generic HDR along d gives as presheaf the pullback

$$\begin{array}{ccc} B \times_{\mathbb{T}_0} \mathbb{T}_1 & \xrightarrow{p_1} & B \\ \downarrow{\scriptstyle p_2} & & \downarrow{\scriptstyle d} \\ \mathbb{T}_1 & \xrightarrow{\mathrm{cod}} & \mathbb{T}_0, \end{array}$$

together with $d' = \mathrm{dom}.p_2$ and ρ' the unique filler of

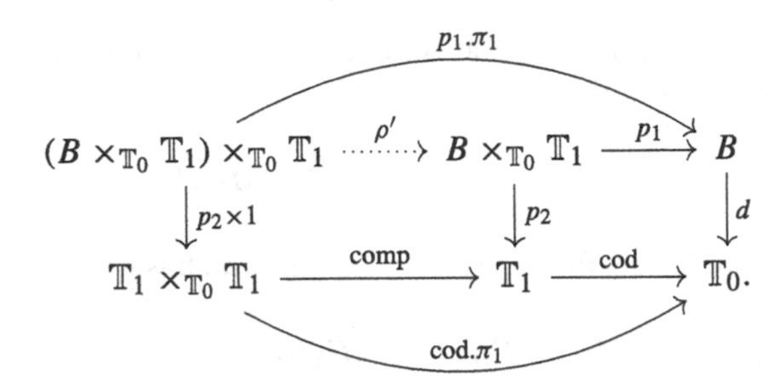

One can interpret this presheaf as follows: the category of internal presheaves has a forgetful functor to the slice category over $\mathbb{T}_0$ and this forgetful functor has a left adjoint. The presheaf $(B \times_{\mathbb{T}_0} \mathbb{T}_1, d', \rho')$ is the free presheaf on $d : B \to \mathbb{T}_0$. Therefore there is an epic morphism of presheaves

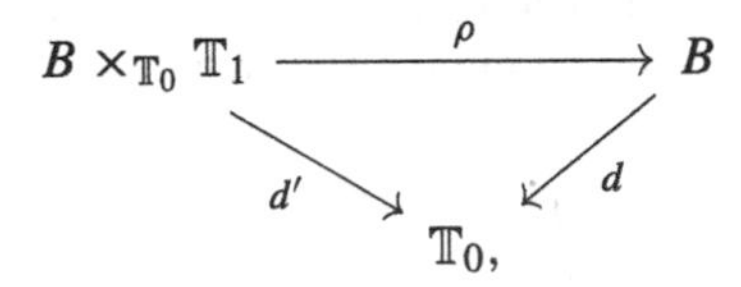

namely the counit of the adjunction. □

10.3 A Small Double Category of Hyperdeformation Retracts

Our next goal is to show that the naive Kan fibrations in simplicial sets are generated by a small double category. We do this by showing that the large double category of HDRs in simplicial sets contains a small double category such that a system of lifts against the small double category can always be extended in a unique way to a system of lifts again the entire double category of HDRs.

Let $\mathbb{H}$ be the following double category.

- Objects are pairs (n, θ) with $n \in \mathbb{N}$ and θ an n-dimensional traversal.
- There is a unique vertical map $(n_0, \theta_0) \to (n_1, \theta_1)$ if $n_0 = n_1$ and θ_0 is a final segment of θ_1.
- A horizontal map $(m, \psi) \to (n, \theta)$ is a pair consisting of a map $\alpha : [m] \to [n]$ together with an m-dimensional traversal σ such that $\psi * \sigma = \theta \cdot \alpha$. The formula for horizontal composition is $(\alpha, \sigma).(\beta, \tau) = (\alpha.\beta, \tau * (\sigma \cdot \beta))$.
- A square is any picture of the form

$$\begin{array}{ccc} (m, \psi_0) & \xrightarrow{(\alpha,\sigma)} & (n, \theta_0) \\ \downarrow & & \downarrow \\ (m, \psi_1) & \xrightarrow{(\alpha,\sigma)} & (n, \theta_1) \end{array}$$

in which the horizontal arrows have the same label.

Our first goal will be to argue that there is a double functor $\mathbb{H} \to \mathrm{HDR}(\widehat{\mathbf{\Delta}})$ which on the level of objects assigns to every (n, θ) the geometric realization of θ. Recall from the previous chapter that the geometric realization of θ is, by definition, the colimit of the following diagram:

$$\begin{array}{ccccccccccc} \Delta^n & & & & \Delta^n & & & \cdots & & & \Delta^n \\ & \searrow^{d_{\theta(1)^s}} & & \swarrow_{d_{\theta(1)^t}} & & \searrow^{d_{\theta(2)^s}} & & \swarrow_{d_{\theta(2)^t}} \quad \searrow^{d_{\theta(k)^s}} & & \swarrow_{d_{\theta(k)^t}} & \\ & & \Delta^{n+1} & & & & \Delta^{n+1} & & \Delta^{n+1} & & \end{array}$$

From now we will denote this colimit simply as θ, rather than as $\widehat{\theta}$. Note that θ comes with two maps from Δ^n, corresponding to the outer maps $\Delta^n \to \Delta^{n+1}$ in this diagram. We will refer to the map $\Delta^n \to \theta$ induced by the inclusion on the left as s_θ and the map of the same shape induced by the inclusion on the right as t_θ.

In fact, s_θ and t_θ occur as pullbacks of id and id*, as follows:

$$\begin{array}{ccccccc} \Delta^n & \xrightarrow{\theta} & \mathbb{T}_0 & & \Delta^n & \xrightarrow{\theta} & \mathbb{T}_0 \\ \downarrow{\scriptstyle s_\theta} & & \downarrow{\scriptstyle \mathrm{id}} & & \downarrow{\scriptstyle t_\theta} & & \downarrow{\scriptstyle \mathrm{id}^*} \\ \theta & \xrightarrow{k_\theta} & \mathbb{T}_1 & & \theta & \xrightarrow{k_\theta} & \mathbb{T}_1 \\ \downarrow{\scriptstyle j_\theta} & & \downarrow{\scriptstyle \mathrm{cod}} & & \downarrow{\scriptstyle j_\theta} & & \downarrow{\scriptstyle \mathrm{cod}} \\ \Delta^n & \xrightarrow{\theta} & \mathbb{T}_0 & & \Delta^n & \xrightarrow{\theta} & \mathbb{T}_0 \end{array}$$

Indeed, using the notation of Corollary 9.1, we may regard s and t as cartesian natural transformations $U \to \widehat{(-)}$ and sections of the cartesian natural transformation $j : \widehat{(-)} \to U$. Also, since id* is the generic HDR, we may regard t_θ as an HDR. Moreover, the picture on the right as well as the naturality squares of the natural transformation t are cartesian morphisms of HDRs.

A typical vertical morphism in $\mathbb{H}$ is of the form $\psi \to \theta * \psi$. Such a morphism we can equip with an HDR-structure, because the top square in

$$\begin{array}{ccc} \Delta^n & \xrightarrow{s_\psi} & \psi \\ \downarrow{\scriptstyle t_\theta} & & \downarrow{\scriptstyle \iota_2} \\ \theta & \xrightarrow{\iota_1} & \theta * \psi \\ \downarrow{\scriptstyle j_\theta} & & \downarrow{\scriptstyle [s_\psi . j_\theta, 1]} \\ \Delta^n & \xrightarrow{s_\psi} & \psi \end{array}$$

is a pushout. Note that because both squares are pullbacks as well, the diagram will become a bicartesian morphism of HDRs.

This explains how we map vertical morphisms to HDRs. Let us now explain where we map the horizontal maps to. Note that the horizontal maps are of the form (α, σ) which we can write as a composition $(\alpha, <>).(1, \sigma)$, so we only need to explain where we map these composites to. However, the map $(1, \sigma) : \psi \to \psi * \sigma$ is just ι_1, while $(1, \alpha) = \widehat{\alpha} : \psi \cdot \alpha \to \psi$ comes from the functoriality of geometric realization (see Corollary 9.1; from now we will also simply write α).

To explain where we map the squares to, we use the same decomposition. Let us first look at a square coming from $(1, \sigma)$:

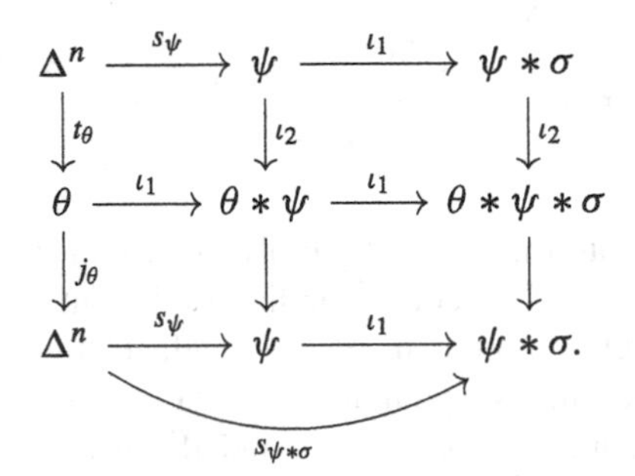

The HDR-structures on the middle and right arrow are defined by pushout from the left arrow: this automatically makes the right hand square a cocartesian morphism of HDRs, and, in particular, a morphism of HDRs. Note that, as before, all squares in the diagram above are pullbacks as well, so the morphism is actually bicartesian.

If the square comes from $(\alpha, <>)$, consider a double cube of the form:

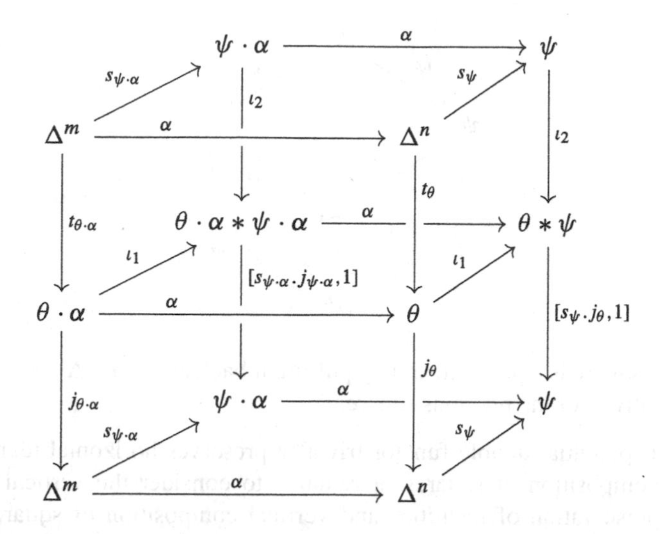

Note that the left and right hand side of this double cube are cocartesian morphisms of HDRs, while the front is a cartesian morphism (since both come with a cartesian morphism to the generic HDR). Since the bottom square is a pullback, the back is a cartesian morphism of HDRs, by Beck-Chevalley (see Proposition 4.6).

This finishes the *construction* of a potential double functor $\mathbb{H} \to \mathrm{HDR}(\widehat{\mathbf{\Delta}})$: the *verification* that it actually is a double functor turns out to be a lot of work and will keep us occupied for the next couple of pages.

Remark 10.1 Note that all the squares in the image of this (potential) double functor are cartesian morphisms of HDRs.

Lemma 10.1 *The potential double functor* $\mathbb{H} \to \mathrm{HDR}(\widehat{\mathbf{\Delta}})$ *just constructed preserves horizontal composition of morphisms.*

Proof To prove that our double functor preserves horizontal composition it suffices to check that

$$\begin{array}{ccc} \psi \cdot \alpha & \xrightarrow{\alpha} & \psi \\ \big\downarrow{\scriptstyle \iota_1} & & \big\downarrow{\scriptstyle \iota_1} \\ \psi \cdot \alpha * \sigma \cdot \alpha & \xrightarrow{\alpha} & \psi * \sigma \end{array}$$

commutes. However, we have a commutative diagram

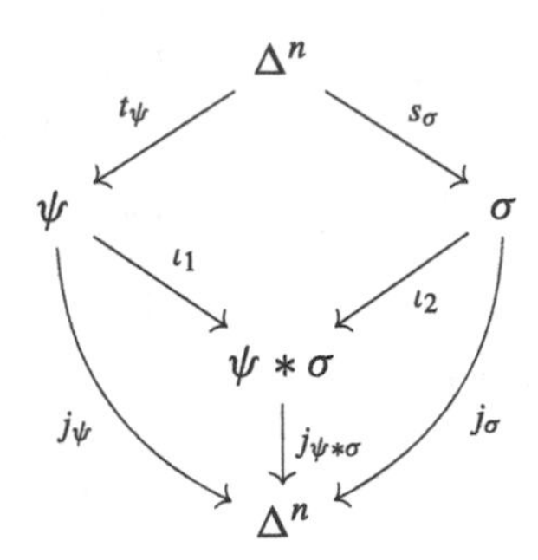

in which the square is a pushout, and by pulling it back along $\alpha : \Delta^m \to \Delta^n$ we get the commutativity of the previous square. □

Since our potential double functor trivially preserves horizontal identities and horizontal composition of squares, it remains to consider the vertical structure. Also here preservation of identities and vertical composition of squares will be immediate once we show vertical composition of morphisms is preserved. For that, it is convenient to use an alternative construction of the HDR-structure on $\psi \to \theta * \psi$, via the "generic inclusion of HDRs". Indeed, consider the following picture:

$$\begin{array}{ccccc}
\mathbb{T}_0 & \xleftarrow{\pi_1} & \mathbb{T}_0 \times \mathbb{T}_0 & \xrightarrow{1\times \mathrm{id}} & \mathbb{T}_0 \times \mathbb{T}_1 \\
\downarrow{\scriptstyle \mathrm{id}^*} & & \downarrow{\scriptstyle \mathrm{id}^*\times 1} & & \downarrow{\scriptstyle \iota_2} \\
\mathbb{T}_1 & \xleftarrow{\pi_1} & \mathbb{T}_1 \times \mathbb{T}_0 & \xrightarrow{\iota_1} & \mathbb{T}_1 \times \mathbb{T}_0 \cup_{\mathbb{T}_0\times\mathbb{T}_0} \mathbb{T}_0 \times \mathbb{T}_1 \\
\downarrow{\scriptstyle \mathrm{cod}} & & \downarrow{\scriptstyle \mathrm{cod}\times 1} & & \downarrow{\scriptstyle [\mathrm{cod}\times\mathrm{id},1]} \\
\mathbb{T}_0 & \xleftarrow{\pi_1} & \mathbb{T}_0 \times \mathbb{T}_0 & \xrightarrow{1\times \mathrm{id}} & \mathbb{T}_0 \times \mathbb{T}_1
\end{array}$$

We can give ι_2 the structure of an HDR, by first pulling the generic structure on id^* back along π_1 and then pushing the result forward along $1 \times \mathrm{id}$. By pulling this HDR back along the horizontal arrow at the centre of

$$\begin{array}{ccc}
\Delta^n & \xrightarrow{(\theta,\psi)} & \mathbb{T}_0 \times \mathbb{T}_0 \\
\downarrow{\scriptstyle s_\psi} & & \downarrow{\scriptstyle 1\times\mathrm{id}} \\
\psi & \longrightarrow & \mathbb{T}_0 \times \mathbb{T}_1 \\
\downarrow{\scriptstyle j_\psi} & & \downarrow{\scriptstyle 1\times\mathrm{cod}} \\
\Delta^n & \xrightarrow{(\theta,\psi)} & \mathbb{T}_0 \times \mathbb{T}_0
\end{array}$$

we obtain $\psi \to \theta * \psi$. The reason is that we can apply Beck-Chevalley to the top square in the diagram above and the HDR $\mathrm{id}^* \times 1$ obtained by pulling the generic HDR along π_1.

Let us translate the generic inclusion in presheaf language. First, the pullback of the generic HDR along $\pi_1 : \mathbb{T}_0 \times \mathbb{T}_0 \to \mathbb{T}_0$ is $\mathbb{T}_1 \times \mathbb{T}_0$ with $d = \text{dom}.\pi_1$ and

$$\rho = (\text{comp}.(\pi_1.\pi_1, \pi_2), \pi_2.\pi_1) : (\mathbb{T}_1 \times \mathbb{T}_0) \times_{\mathbb{T}_0} \mathbb{T}_1 \to \mathbb{T}_1 \times \mathbb{T}_0.$$

Then we need to push this forward along $1 \times \text{id} : \mathbb{T}_0 \times \mathbb{T}_0 \to \mathbb{T}_0 \times \mathbb{T}_1$, which results in:

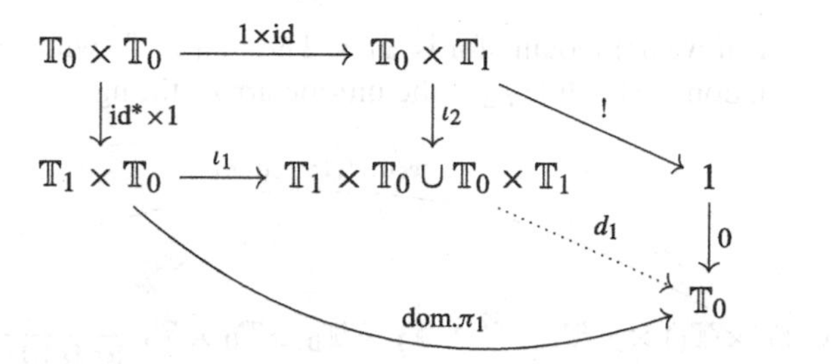

So the generic inclusion of HDRs is $\mathbb{T}_1 \times \mathbb{T}_0 \cup \mathbb{T}_0 \times \mathbb{T}_1$ with $d_1 = [\text{dom}.\pi_1, 0.!]$. It also comes equipped with an action ρ_1, which we find as follows:

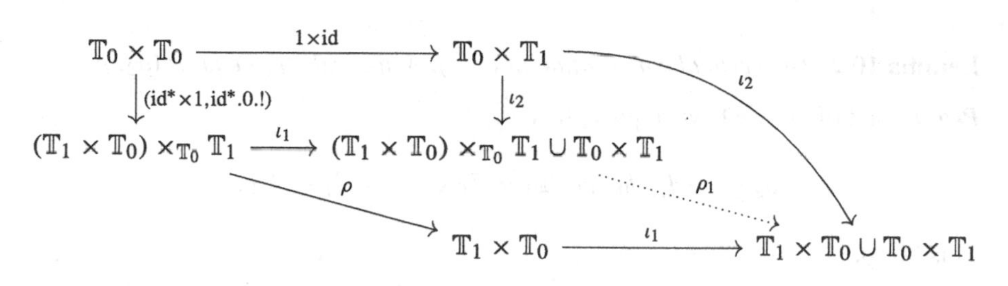

Hence $\rho_1 = (\text{comp}.(\pi_1.\pi_1, \pi_2), \pi_2.\pi_1) \cup 1$.

We will compare this with two other HDRs: the pullbacks of the generic HDR along the maps $\pi_2, * : \mathbb{T}_0 \times \mathbb{T}_0 \to \mathbb{T}_0$. In terms of presheaves, the pullback of the generic HDR along $\pi_2 : \mathbb{T}_0 \times \mathbb{T}_0 \to \mathbb{T}_0$ is $d_0 = \text{dom}.\pi_2 : \mathbb{T}_0 \times \mathbb{T}_1 \to \mathbb{T}_0$, while ρ_0 is the unique map filling

$$\begin{array}{ccccc} \mathbb{T}_0 \times (\mathbb{T}_1 \times_{\mathbb{T}_0} \mathbb{T}_1) & \xrightarrow{\rho_0} & \mathbb{T}_0 \times \mathbb{T}_1 & \xrightarrow{1\times\text{cod}} & \mathbb{T}_0 \times \mathbb{T}_0 \\ \downarrow{\scriptstyle \pi_2} & & \downarrow{\scriptstyle \pi_2} & & \downarrow{\scriptstyle \pi_2} \\ \mathbb{T}_1 \times_{\mathbb{T}_0} \mathbb{T}_1 & \xrightarrow{\text{comp}} & \mathbb{T}_1 & \xrightarrow{\text{cod}} & \mathbb{T}_0. \end{array}$$

(with the outer arrow $(\pi_1, \text{cod}.\pi_1.\pi_2) : \mathbb{T}_0 \times (\mathbb{T}_1 \times_{\mathbb{T}_0} \mathbb{T}_1) \to \mathbb{T}_0 \times \mathbb{T}_0$)

Hence $\rho_0 = 1 \times \text{comp}$.

To compute the pullback of the generic HDR along $* : \mathbb{T}_0 \times \mathbb{T}_0 \to \mathbb{T}_0$, we first compute the pullbacks

$$
\begin{array}{ccc}
\mathbb{T}_0 \times \mathbb{T}_0 & \xrightarrow{\quad * \quad} & \mathbb{T}_0 \\
\downarrow{\scriptstyle \iota_2.(1\times \mathrm{id}^*)} & & \downarrow{\scriptstyle \mathrm{id}^*} \\
\mathbb{T}_1 \times \mathbb{T}_0 \cup \mathbb{T}_0 \times \mathbb{T}_1 & \xrightarrow{\quad \mu^* \quad} & \mathbb{T}_1 \\
\downarrow{\scriptstyle [\mathrm{cod}\times 1, 1\times \mathrm{cod}]} & & \downarrow{\scriptstyle \mathrm{cod}} \\
\mathbb{T}_0 \times \mathbb{T}_0 & \xrightarrow{\quad * \quad} & \mathbb{T}_0
\end{array}
$$

Hence the presheaf we are looking for is $\mathbb{T}_1 \times \mathbb{T}_0 \cup \mathbb{T}_0 \times \mathbb{T}_1$ with $d_2 = \mathrm{dom}.\mu^* = [*.(\mathrm{dom}.\pi_1, \pi_2), \mathrm{dom}.\pi_2]$, while ρ_2 is the unique arrow filling

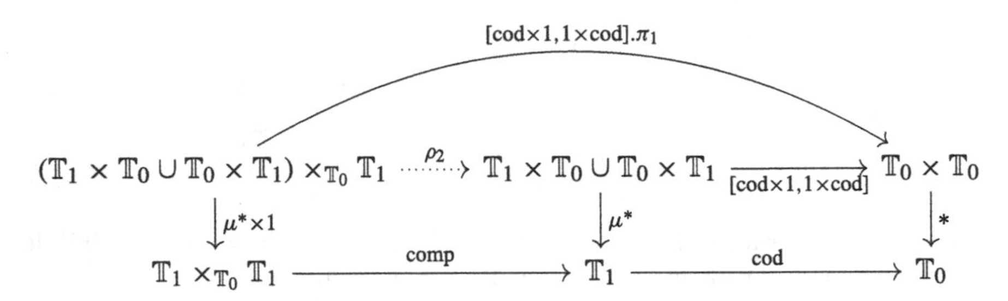

Lemma 10.2 *The vertical composition of* (d_1, ρ_1) *after* (d_0, ρ_0) *equals* (d_2, ρ_2).

Proof First of all, we have to prove that

$$d_2 = *.(d_1, d_0.j_1) : \mathbb{T}_1 \times \mathbb{T}_0 \cup \mathbb{T}_0 \times \mathbb{T}_1 \to \mathbb{T}_0.$$

This holds, because:

$$
\begin{aligned}
*.(d_1, d_0.j_1).\iota_1 &= *.(d_1.\iota_1, d_0.j_1.\iota_1) \\
&= *.(\mathrm{dom}.\pi_1, \mathrm{dom}.\pi_2.(\mathrm{cod} \times \mathrm{id})) \\
&= *.(\mathrm{dom}.\pi_1, \mathrm{dom}.\mathrm{id}.\pi_2) \\
&= *.(\mathrm{dom}.\pi_1, \pi_2) \\
&= d_2.\iota_1
\end{aligned}
$$

and

$$
\begin{aligned}
*.(d_1, d_0.j_1).\iota_2 &= *.(d_1.\iota_2, d_0) \\
&= *.(0.!, \mathrm{dom}.\pi_2) \\
&= \mathrm{dom}.\pi_2 \\
&= d_2.\iota_2.
\end{aligned}
$$

Writing $C := \mathbb{T}_1 \times \mathbb{T}_0 \cup \mathbb{T}_0 \times \mathbb{T}_1$, this means that we can also think of the domain of ρ_2 as the pullback

$$\begin{array}{ccc} C \times_{\mathbb{T}_0} \mathbb{T}_1 \cup C \times_{\mathbb{T}_0} \mathbb{T}_1 & \xrightarrow{\quad q \quad} & C \\ \downarrow{\scriptstyle p} & & \downarrow{\scriptstyle (d_1, d_0.j_1)} \\ \mathbb{T}_1 \times \mathbb{T}_0 \cup \mathbb{T}_0 \times \mathbb{T}_1 & \xrightarrow{[\mathrm{cod}\times 1, 1\times \mathrm{cod}]} & \mathbb{T}_0 \times \mathbb{T}_0 \\ \downarrow{\scriptstyle \mu^*} & & \downarrow{\scriptstyle *} \\ \mathbb{T}_1 & \xrightarrow{\quad \mathrm{cod} \quad} & \mathbb{T}_0. \end{array}$$

where

$$q := [\pi_1, \pi_1],$$
$$p := (\pi_2, d_0.j_1.\pi_1) \cup (d_1 \times 1).$$

Therefore the domain of ρ_2 can be seen as the pushout of two pullbacks, the pullback of $d_1 = [\mathrm{dom}.\pi_1, 0.!] : C \to \mathbb{T}_0$ and cod as well as the pullback of

$$d_0.j_1 = \mathrm{dom}.\pi_2.[\mathrm{cod} \times \mathrm{id}, 1] = [\mathrm{dom.id}.\pi_2, \mathrm{dom}.\pi_2] = [\pi_2, \mathrm{dom}.\pi_2] : C \to \mathbb{T}_0$$

and cod. In these terms, ρ_2 is the unique arrow making

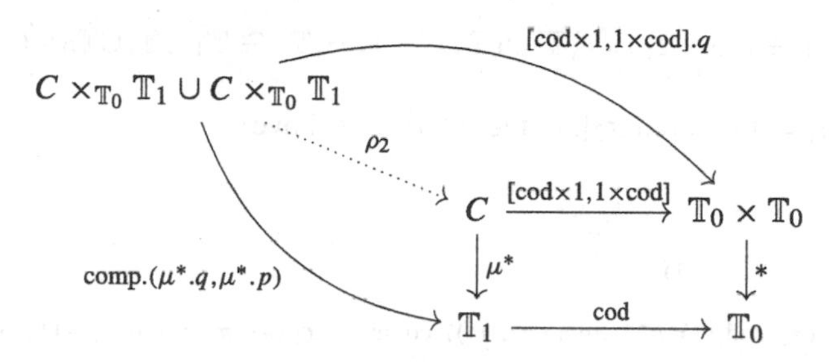

commute.

We have to prove $\rho_2 = [\rho_1, i_1.\rho_0.(j_1 \times 1)]$. Our strategy will be to prove that these morphisms agree on both summands, and that in turn we will do by showing that on these summands the maps become equal upon postcomposing with both μ^* and $[\mathrm{cod} \times 1, 1 \times \mathrm{cod}]$. Note that the first summand $C\,{}^{d_1}\!\times^{\mathrm{cod}}_{\mathbb{T}_0} \mathbb{T}_1$ is isomorphic to

$$[(\mathbb{T}_1 \times \mathbb{T}_0)\,{}^{\mathrm{dom}.\pi_1}\!\times^{\mathrm{cod}}_{\mathbb{T}_0} \mathbb{T}_1] \cup [\mathbb{T}_0 \times \mathbb{T}_1],$$

and in these terms we have

$$
\begin{aligned}
&\mathrm{comp}.(\mu^*.q, \mu^*.p).\iota_1 &=\\
&\mathrm{comp}.(\mu^*.\pi_1, \mu^*.\iota_1.(\pi_2, d_0.j_1.\pi_1)) &=\\
&[\mathrm{comp}.(\mu^*.\iota_1.\pi_1, \mu^*.\iota_1.(\pi_2, \pi_2.\pi_1)), \mathrm{comp}.(\mu^*.\iota_2, \mu^*.\iota_1.(\mathrm{id}^*.0.!, \mathrm{dom}.\pi_2))] &=\\
&[\mu^*.\iota_1.(\mathrm{comp}.(\pi_1.\pi_1, \pi_2), \pi_2.\pi_1), \mu^*.\iota_2] &=\\
&\mu^*.\rho_1
\end{aligned}
$$

as well as

$$
\begin{aligned}
&[\mathrm{cod} \times 1, 1 \times \mathrm{cod}].q.\iota_1 &=\\
&[\mathrm{cod} \times 1, 1 \times \mathrm{cod}].\pi_1 &=\\
&[\mathrm{cod} \times 1.\pi_1, 1 \times \mathrm{cod}] &=\\
&[(\mathrm{cod}.\pi_1.\pi_1, \pi_2.\pi_1), 1 \times \mathrm{cod}] &=\\
&[\mathrm{cod} \times 1.(\mathrm{comp}.(\pi_1.\pi_1, \pi_2), \pi_2.\pi_1), 1 \times \mathrm{cod}] &=\\
&[\mathrm{cod} \times 1, 1 \times \mathrm{cod}].\rho_1.
\end{aligned}
$$

We now turn to the second summand, which we can write as

$$
C^{d_0.j_1} \times_{\mathbb{T}_0}^{\mathrm{cod}} \mathbb{T}_1 \cong (\mathbb{T}_1 \times \mathbb{T}_0) \times_{\mathbb{T}_0} \mathbb{T}_1 \cup (\mathbb{T}_0 \times \mathbb{T}_1) \times_{\mathbb{T}_0} \mathbb{T}_1 \cong \mathbb{T}_1 \times \mathbb{T}_1 \cup \mathbb{T}_0 \times (\mathbb{T}_1 \times_{\mathbb{T}_0} \mathbb{T}_1),
$$

because $d_0.j_1 = [\pi_2, \mathrm{dom}.\pi_2]$. In these terms we have

$$
\begin{aligned}
&\mathrm{comp}.(\mu^*.q, \mu^*.p).\iota_2 &=\\
&\mathrm{comp}.(\mu^*.\pi_1, \mu^*.\iota_2.d_1 \times 1) &=\\
&[\mathrm{comp}.(\mu^*.\iota_1.(\pi_1, \mathrm{cod}.\pi_2), \mu^*.\iota_2.(\mathrm{dom}.\pi_1, \pi_2)), \mathrm{comp}.(\mu^*.\iota_2.(\pi_1, \pi_1.\pi_2), \mu^*.\iota_2.(0.!, \pi_2.\pi_2))] &=\\
&[\mu^*.\iota_2.(\mathrm{cod}.\pi_1, \pi_2), \mu^*.\iota_2.1 \times \mathrm{comp}] &=
\end{aligned}
$$

which equals:

$$
\begin{aligned}
&\mu^*.i_1.\rho_0.(j_1 \times 1) &=\\
&\mu^*.\iota_2.(1 \times \mathrm{comp}).([\mathrm{cod} \times \mathrm{id}, 1] \times 1) &=\\
&\mu^*.\iota_2.[(\mathrm{cod}.\pi_1, \mathrm{comp}.(\mathrm{id}.\mathrm{cod}.\pi_2, \pi_2)), 1 \times \mathrm{comp}] &=\\
&\mu^*.\iota_2.[(\mathrm{cod}.\pi_1, \pi_2), 1 \times \mathrm{comp}].
\end{aligned}
$$

In a similar fashion we have:

$$\begin{aligned}
&[\mathrm{cod} \times 1, 1 \times \mathrm{cod}].i_1.\rho_0.(j_1 \times 1) &=\\
&[\mathrm{cod} \times 1, 1 \times \mathrm{cod}].\iota_2.(1 \times \mathrm{comp}).([\mathrm{cod} \times \mathrm{id}, 1] \times 1) &=\\
&1 \times \mathrm{cod}.[(\mathrm{cod}.\pi_1, \pi_2), 1 \times \mathrm{comp}] &=\\
&[(\mathrm{cod}.\pi_1, \mathrm{cod}.\pi_2), (\pi_1, \mathrm{cod}.\pi_1.\pi_2)] &=\\
&[\mathrm{cod} \times 1, 1 \times \mathrm{cod}].\pi_1 &=\\
&[\mathrm{cod} \times 1, 1 \times \mathrm{cod}].q.\iota_2.
\end{aligned}$$

This finishes the proof. □

Remark 10.2 The previous lemma is equivalent to the distributive law in the form:

$$\Gamma.\mu = M\mu.(\mu M.(\Gamma.p_1, Mr.p_2), \mu M.(\alpha.(p_2, M!.p_1), \Gamma.p_2)).$$

To prove that this reformulation is equivalent to the way we usually state the distributive law, one would need an interchange law of the following form: suppose we have elements $\alpha, \beta, \gamma, \delta \in MMX$ with $t(\alpha) = s(\beta), t(\gamma) = s(\delta), Mt(\alpha) = Ms(\gamma), Mt(\beta) = Ms(\gamma)$, then:

$$M\mu.(\mu M(\alpha, \beta), \mu M(\gamma, \delta)) = \mu M.(M\mu.(\alpha, \gamma), M\mu.(\beta, \delta)).$$

This interchange law may not hold in all Moore categories, but it can be shown to hold in the example at hand.

In any case, to see the equivalence between the previous lemma and the reformulation of the distributive law, one would have to think about how one would prove the latter, and for that we have to go back to the proof of Lemma 9.1. The description of $\Gamma.\mu$ is still correct of course, but note that in the negative direction we can think of it as a map $C \times_{\mathbb{T}_0} \mathbb{T}_1 \to C$:

$$\begin{array}{ccccc}
C & \longleftarrow & C \times_{\mathbb{T}_0} \mathbb{T}_1 & \longrightarrow & \{(\theta \in \mathbb{T}_0, t : \mathrm{cod}^{-1}(\theta) \to \mathbb{T}_0, p_0 \in \theta, p_1 \in t(p_0))\} \\
& \searrow & \downarrow & & \downarrow \\
& & \mathbb{T}_0 \times \mathbb{T}_0 & \longrightarrow & \{(\theta, t : \mathrm{cod}^{-1}(\theta) \to \mathbb{T}_0)\}
\end{array}$$

where the arrow along the bottom sends (θ_0, θ_1) to $(\theta_0 * \theta_1, d_2 = d_1.\mu^* : C \to \mathbb{T}_0)$. With some effort one can recognise the map $C \times_{\mathbb{T}_0} \mathbb{T}_1 \to C$ as ρ_2.

Let us first look at the map $\mu M.(\Gamma.p_1, Mr.p_2)$, which can be thought of as a composite:

$$\begin{array}{ccc} \mathrm{cod} \times_I \mathrm{cod} & \longrightarrow & (\mathrm{cod} \otimes \mathrm{cod}) \times_{I \otimes \mathrm{cod}} (\mathrm{cod} \otimes \mathrm{cod}) \\ & & \downarrow \cong \\ & & (\mathrm{cod} \times_I \mathrm{cod}) \otimes \mathrm{cod} \xrightarrow{\mu \otimes 1_{\mathrm{cod}}} \mathrm{cod} \otimes \mathrm{cod}. \end{array}$$

In the positive direction this sends (θ_0, θ_1) first to $((\theta_0, \lambda p.\mathrm{dom}(p), (\theta_1, \lambda p.0))$ and then to what is essentially $(\theta_0 * \theta_1, d_1 = [\mathrm{dom}.\pi_1, 0.!] : C \to \mathbb{T}_0)$. Then in the negative direction we have a map $C^{d_1} \times^{\mathrm{cod}}_{\mathbb{T}_0} \mathbb{T}_1 \to C$ which with some effort one can recognise as ρ_1.

Let us now have a look at $\mu M.(\alpha.(p_2, M!.p_1), \Gamma.p_2)$. We can think of it as a morphism

$$\begin{array}{ccc} \mathrm{cod} \times_I \mathrm{cod} & \longrightarrow & (\mathrm{cod} \otimes \mathrm{cod}) \times_{I \otimes \mathrm{cod}} (\mathrm{cod} \otimes \mathrm{cod}) \\ & & \downarrow \cong \\ & & (\mathrm{cod} \times_I \mathrm{cod}) \otimes \mathrm{cod} \xrightarrow{\mu \otimes 1_{\mathrm{cod}}} \mathrm{cod} \otimes \mathrm{cod}. \end{array}$$

In the positive direction this sends a pair (θ_0, θ_1) first to $((\theta_0, \lambda p.\theta_1), (\theta_1, \lambda p.\mathrm{dom}(p))$ which then gets sent to what is essentially $(\theta_0 * \theta_1, d_0.j_1 = [\pi_2, \mathrm{dom}.\pi_2] : C \to \mathbb{T}_0)$. This means that in the negative direction we should have a map $C^{d_0.j_1} \times^{\mathrm{cod}}_{\mathbb{T}_0} \mathbb{T}_1 \to C$: again with some effort one can recognise this as $i_1.\rho_0.(j_1 \times 1)$.

Finally, the total right hand side is a map

$$\begin{array}{ccc} \mathrm{cod} \times_I \mathrm{cod} & \longrightarrow & (\mathrm{cod} \otimes \mathrm{cod}) \times_{\mathrm{cod} \otimes I} (\mathrm{cod} \otimes \mathrm{cod}) \\ & & \downarrow \cong \\ & & \mathrm{cod} \otimes (\mathrm{cod} \times_I \mathrm{cod}) \xrightarrow{1_{\mathrm{cod}} \otimes \mu} \mathrm{cod} \otimes \mathrm{cod}. \end{array}$$

In the positive direction this will be a map sending (θ_0, θ_1) to $(\theta_0 * \theta_1, *.(d_1, j_0.d_0) : C \to \mathbb{T}_0)$; in the backwards direction this is a map $C \times_{\mathbb{T}_0} \mathbb{T}_1 \cup C \times_{\mathbb{T}_0} \mathbb{T}_1 \to C$ which is $[\rho_1, i_1.\rho_0.(j_1 \times 1)]$. This shows that the reformulated distributive law is equivalent to the previous lemma.

Proposition 10.3 *There is a double functor* $\mathbb{H} \to \mathrm{HDR}(\widehat{\mathbf{\Delta}})$ *which on the level of objects assigns to every traversal its geometric realization.*

Proof As said, it remains to check that vertical composition of morphisms is preserved. We start by pulling back the vertical composition from the previous lemma along $(\theta, \psi) : \Delta^n \to \mathbb{T}_0 \times \mathbb{T}_0$ and obtain the following picture:

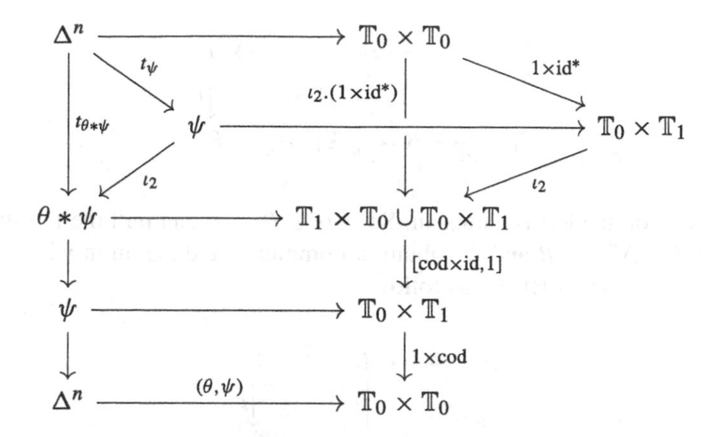

Since vertical composition is preserved by pullback (see Part I) and we have a commutative triangle of HDRs on the right, the same is true on the left. Since push forward also preserves vertical composition, we can push this triangle forward along $s_\psi : \Delta^n \to \psi$ to see that our double functor does indeed preserve vertical composition. □

This lengthy verification was only the first step towards proving the main result of this section, which is:

Theorem 10.2 *The following notions of fibred structure are isomorphic:*

- *Having the right lifting property against the large double category of HDRs in simplicial sets (that is, to be a naive Kan fibration).*
- *Having the right lifting property against the small double category $\mathbb{H}$.*

In addition, the morphism of discretely fibred concrete double categories $HDR^{\pitchfork} \to \mathbb{H}^{\pitchfork}$ induced by the inclusion of $\mathbb{H}$ into the double category of HDRs in simplicial sets satisfies fullness on squares and is therefore an isomorphism.

Proof First, we explain how one can construct a morphism of notions of fibred structure in the other direction. So imagine that we have a map $p : Y \to X$ which has the right lifting property against $\mathbb{H}$ and we are given a lifting problem

$$\begin{array}{ccc} A & \xrightarrow{g} & Y \\ {\scriptstyle i}\downarrow & \overset{l}{\nearrow} & \downarrow{\scriptstyle p} \\ B & \xrightarrow{f} & X \end{array}$$

where $(i : A \to B, j, H)$ is an HDR.

In the proof of the existence of a generic HDR, we have seen that for any HDR (i, j, H) there are two morphisms of HDRs, as in

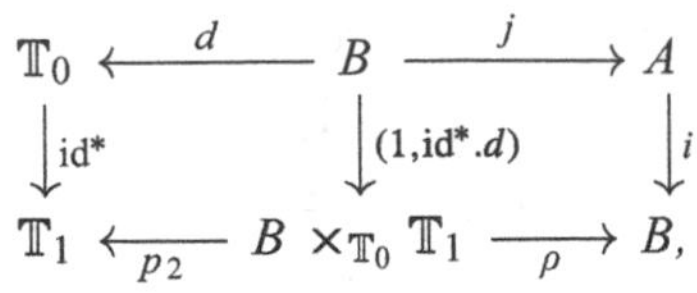

$$\begin{array}{ccccc} \mathbb{T}_0 & \xleftarrow{d} & B & \xrightarrow{j} & A \\ \downarrow{\scriptstyle \mathrm{id}^*} & & \downarrow{\scriptstyle (1,\mathrm{id}^*.d)} & & \downarrow{\scriptstyle i} \\ \mathbb{T}_1 & \xleftarrow[p_2]{} & B \times_{\mathbb{T}_0} \mathbb{T}_1 & \xrightarrow[\rho]{} & B, \end{array}$$

where the one on the left is cartesian. So if $b \in B_n$, we can pull the middle HDR back along $b : \Delta^n \to B$ and we obtain a commutative diagram in which the left square is a morphism of HDRs, as follows:

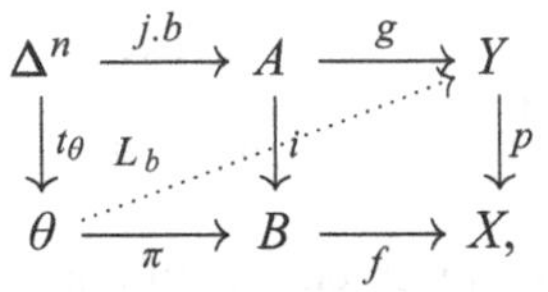

$$\begin{array}{ccccc} \Delta^n & \xrightarrow{j.b} & A & \xrightarrow{g} & Y \\ \downarrow{\scriptstyle t_\theta} & L_b & \downarrow{\scriptstyle i} & & \downarrow{\scriptstyle p} \\ \theta & \xrightarrow[\pi]{} & B & \xrightarrow[f]{} & X, \end{array}$$

if $d(b) = \theta$. By assumption, we have a dotted filler L_b, as indicated. Now we put $l(b) := L_b.s_\theta$.

Let us first check that this defines a natural transformation $l : B \to Y$. If we consider $b \cdot \alpha$ for some $\alpha : \Delta^m \to \Delta^n$, then we have a picture as follows:

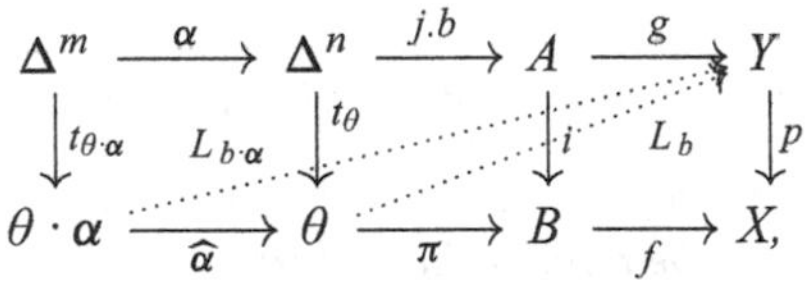

$$\begin{array}{ccccccc} \Delta^m & \xrightarrow{\alpha} & \Delta^n & \xrightarrow{j.b} & A & \xrightarrow{g} & Y \\ \downarrow{\scriptstyle t_{\theta\cdot\alpha}} & L_{b\cdot\alpha} & \downarrow{\scriptstyle t_\theta} & & \downarrow{\scriptstyle i} & L_b & \downarrow{\scriptstyle p} \\ \theta\cdot\alpha & \xrightarrow[\widehat{\alpha}]{} & \theta & \xrightarrow[\pi]{} & B & \xrightarrow[f]{} & X, \end{array}$$

Hence

$$l(b \cdot \alpha) = L_{b\cdot\alpha}.s_{\theta\cdot\alpha} = L_b.\widehat{\alpha}.s_{\theta\cdot\alpha} = L_b.s_\theta.\alpha = l(b).\alpha,$$

which shows that l is indeed a natural transformation.

Let us now check that l fills the original square. Because $\pi.s_\theta = b$, we have

$$p.l(b) = p.L_b.s_\theta = f.\pi.s_\theta = f(b),$$

hence l makes the lower triangle commute. Also, if $b = i(a)$ for some $a \in A_n$, then $d(b) = \langle\rangle$ (the empty traversal), and

$$l.i(a) = L_b.s_{\langle\rangle} = L_b.t_{\langle\rangle} = g.j(b) = g.j.i(a) = g(a).$$

It remains to check that these lifts satisfy both the horizontal and vertical compatibility conditions. We start by looking at the horizontal one. Imagine we have a commutative diagram of the form:

$$\begin{array}{ccccc} A' & \xrightarrow{\alpha} & A & \longrightarrow & Y \\ \downarrow{\scriptstyle i'} & l' & \downarrow{\scriptstyle i} & l & \downarrow{\scriptstyle p} \\ B' & \xrightarrow[\beta]{} & B & \longrightarrow & X \end{array}$$

where the square on the left is a morphism of HDRs and $b' \in B'_n$. Then this fits into a larger picture:

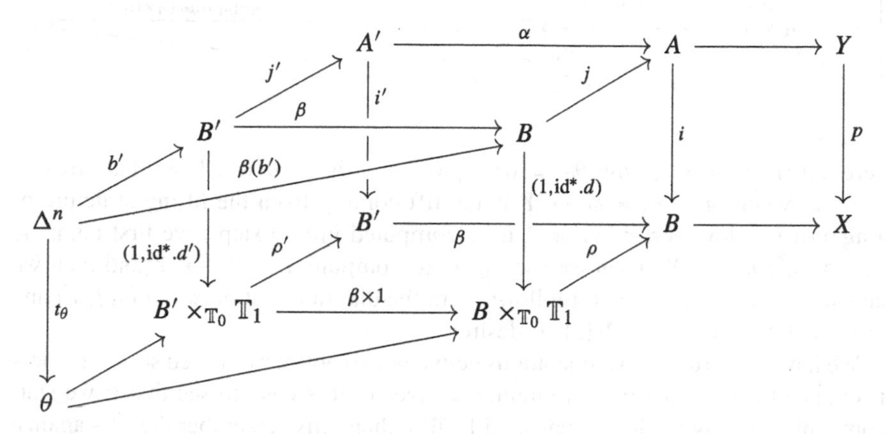

with $d' = d.\beta$. So from left to right we obtain a lift $\pi : \theta \to Y$ and $l'(b') = \pi.s_\theta$. But because the front face of the cube is a cartesian morphism of HDRs (as one easily checks), $l(\beta(b'))$ is computed in the same way. This shows the horizontal compatibility condition.

For checking the vertical compatibility condition, imagine that we have a lifting problem of the form:

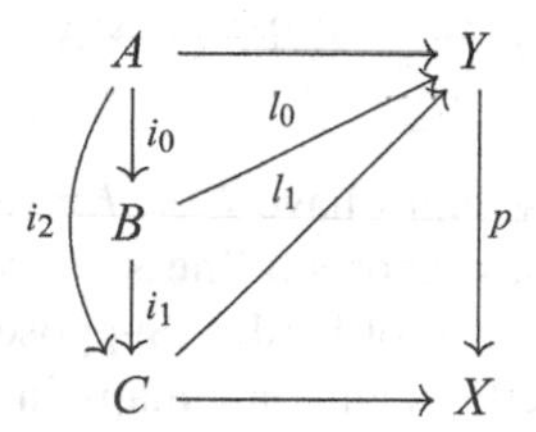

with the HDR i_2 the vertical composition of i_0 and i_1, l_0 and l_1 the lifts induced by i_0 and i_1, respectively, and $c \in C_n$. The aim is to show that for the induced lift $l_2 : C \to Y$ induced by i_2 we have $l_2(c) = l_1(c)$. Using the formulas for the vertical composition of HDRs (in the language of presheaves) we obtain a commutative diagram, as follows:

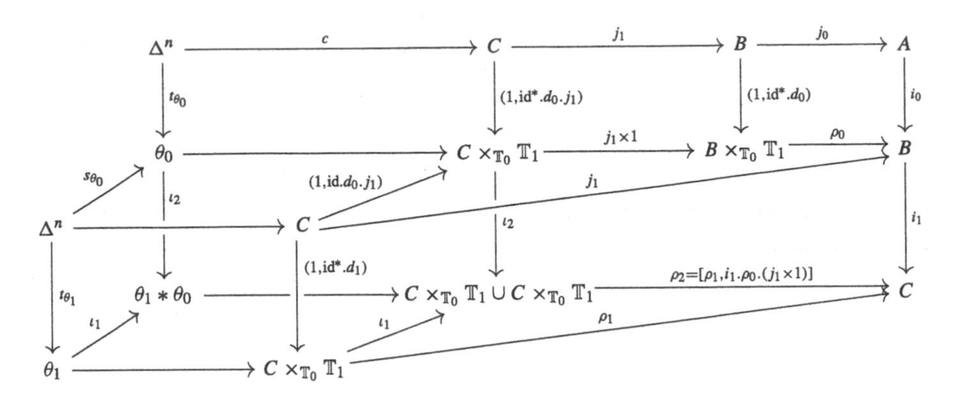

Here $d_2(c) = \theta_1 * \theta_0$ with $\theta_0 = (d_0.j_1)(c)$ and $\theta_1 = d_1(c)$. Note that $l_2(c) = \pi_2.s_{\theta_1 * \theta_0}$ where $\pi_2 : \theta_1 * \theta_0 \to Y$ is the lift coming from the lifting structure of p against $\mathbb{H}$. However, $\pi_2.s_{\theta_1 * \theta_0}$ can be computed in two steps: we first compute the lift $\pi_0 : \theta_0 \to Y$. Then we use $\pi_0.s_{\theta_0}$ to compute $\pi_1 : \theta_1 \to Y$ and then we have $\pi_2.s_{\theta_1 * \theta_0} = \pi_1.s_{\theta_1}$. But it follows from the diagram that $\pi_0.s_{\theta_0} = l_0(j_1.c)$ and $\pi_1.s_{\theta_1} = l_1(c)$: so $l_2(c) = l_1(c)$, as desired.

We have constructed two operations between two notions of fibred structure: now it remains to show that they are mutually inverse. It is easy to see that if we start from a map having the RLP against all HDRs, then only remember the lifts against the vertical maps in $\mathbb{H}$ and then use the operation defined above to compute a lift against a general HDR, we return at our starting point. The reason is simply that the left hand square in

$$\begin{array}{ccccc} \Delta^n & \xrightarrow{j.b} & A & \xrightarrow{g} & Y \\ \downarrow{\scriptstyle t_\theta} & L & \downarrow{\scriptstyle i} & l & \downarrow{\scriptstyle p} \\ \theta & \xrightarrow[\pi]{} & B & \xrightarrow[f]{} & X, \end{array}$$

is a morphism of HDRs, so we must have $L = l.\pi$. So if $b = \pi.s_\theta$, then $L.s_\theta = l.\pi.s_\theta = l(b)$. This argument also shows fullness on squares.

The converse turns out to be a lot harder. Suppose that we start with a map p having the right lifting property against all maps in $\mathbb{H}$, and that we are given a lifting problem of p against a vertical map from $\mathbb{H}$. Now we can find a solution in two different ways: first, we can use the lifting structure of p directly. Alternatively, we can use that vertical maps in $\mathbb{H}$ are HDRs and use this to find a lift, following the procedure explained above. The question is: are both solutions necessarily the same? We claim that the answer is yes.

To prove the claim, it suffices to check that the lifts against traversals of length 1 are identical. The reason is that any inclusion of traversals can be written as the vertical composition of inclusions where each next traversal has length one longer than the previous. And if we have a traversal of the form $\sigma \to < (i, \pm) > * \sigma$, then there is a bicartesian square in $\mathbb{H}$ of the form:

$$\begin{array}{ccc} <> & \longrightarrow & \sigma \\ \downarrow & & \downarrow \\ <(i,\pm)> & \longrightarrow & <(i,\pm)> * \sigma \end{array}$$

So the lift against the map on the right is completely determined by the lift against the map on the left.

So let us imagine that we have an inclusion of traversals of the form $<> \to <(i,+)>$ (we will only look at the positive case, for simplicity). Then its geometric realisation is:

$$\begin{array}{ccc} \Delta^n & \xrightarrow{<(i,+)>} & \mathbb{T}_0 \\ {\scriptstyle d_i}\downarrow & & \downarrow{\scriptstyle \mathrm{id}^*} \\ \Delta^{n+1} & \xrightarrow{u} & \mathbb{T}_1 \\ {\scriptstyle s_i}\downarrow & & \downarrow{\scriptstyle \mathrm{cod}} \\ \Delta^n & \xrightarrow{<(i,+)>} & \mathbb{T}_0 \end{array}$$

where $u = k_{<i,+>}$ picks out the traversal $<(i,+)> \cdot s_i =< (i+1,+), (i,+) >$ with the special position (so the position in the middle). Note that this means that face maps $d_i : \Delta^n \to \Delta^{n+1}$ are HDRs: let us see what its HDR-structure is in presheaf language. First of all, we have $d = \mathrm{dom}.u =< (i,+) >: \Delta^{n+1} \to \mathbb{T}_0$. Secondly, we have to determine $\rho : \Delta^{n+1} \times_{\mathbb{T}_0} \mathbb{T}_1 \to \Delta^{n+1}$. But note that the domain of ρ also arises as the centre left object in

$$\begin{array}{ccc} \Delta^{n+1} & \xrightarrow{d} & \mathbb{T}_0 \\ {\scriptstyle d_i}\downarrow & & \downarrow{\scriptstyle \mathrm{id}^*} \\ \Delta^{n+2} & \xrightarrow{v} & \mathbb{T}_1 \\ {\scriptstyle s_i}\downarrow & & \downarrow{\scriptstyle \mathrm{cod}} \\ \Delta^{n+1} & \xrightarrow[d=<(i,+)>]{} & \mathbb{T}_0 \end{array}$$

where v chooses $<(i,+)> \cdot s_i$ with the special position; in other words, it is isomorphic to Δ^{n+2}. This means that ρ is the unique map filling

$$\begin{array}{ccccc} \Delta^{n+2} & \overset{\rho}{\dashrightarrow} & \Delta^{n+1} & \xrightarrow{s_i} & \Delta^n \\ {\scriptstyle (u \cdot s_i, v)}\downarrow & \searrow^{q} & \downarrow{\scriptstyle u} & & \downarrow{\scriptstyle (<i,+>)} \\ \mathbb{T}_1 \times_{\mathbb{T}_0} \mathbb{T}_1 & \xrightarrow[\mathrm{comp}]{} & \mathbb{T}_1 & \xrightarrow[\mathrm{cod}]{} & \mathbb{T}_0. \end{array}$$

(with the curved top arrow $s_i . s_i : \Delta^{n+2} \to \Delta^n$)

Note that $u \cdot s_i$ is $< (i+2,+), (i+1,+), (i,+) >$ with the position between the first and second element, and therefore q is $< (i+2,+), (i+1,+), (i,+) >$ with the position between the second and third elements. We conclude that ρ must be s_{i+1}.

What we have to prove, then, is that if we have a lifting problem of the form

$$\begin{array}{ccc} \Delta^n & \longrightarrow & Y \\ \Big\downarrow{\scriptstyle d_i} & \nearrow & \Big\downarrow{\scriptstyle p} \\ \Delta^{n+1} & \longrightarrow & X \end{array}$$

where we think of the arrow on the left as an HDR coming from the inclusion of traversals $<> \to < (i,+) >$, then the lift coming from the fact that it is an HDR coincides with the one coming from the fact that p has the right lifting property against $\mathbb{H}$. The former lift is computed by choosing an arbitrary element $\alpha \in \Delta^{n+1}$ and pulling the map on the left back along $s_i.\alpha$. But because Δ^{n+1} is representable it suffices to do this for $\alpha = 1$, which means that what we have to prove is that if we have a picture as follows:

$$\begin{array}{ccccc} \Delta^{n+1} & \xrightarrow{s_i} & \Delta^n & \longrightarrow & Y \\ \Big\downarrow{\scriptstyle d_i} & \overset{l_2}{\nearrow} & \Big\downarrow{\scriptstyle d_i} & \overset{l_1}{\nearrow} & \Big\downarrow{\scriptstyle p} \\ \Delta^{n+2} & \xrightarrow[s_{i+1}]{} & \Delta^{n+1} & \longrightarrow & X \end{array}$$

then for the lifts coming from the fact that p has the right lifting property against $\mathbb{H}$, we have $l_1 = l_2.d_{i+1}$. For that it suffices to prove $l_1.s_{i+1} = l_2$. (Just to be clear, the left and centre vertical arrows are to be thought of as HDRs coming from the inclusion of traversals $<> \to < (i,+) >$ in dimensions $n+1$ and n, respectively.)

To that end, note that we have a commutative diagram of the form:

$$\begin{array}{ccccc} \Delta^{n+1} & \xrightarrow{s_i} & \Delta^n & \longrightarrow & Y \\ \Big\downarrow{\scriptstyle d_i} & & \Big| & & \Big| \\ < (i,+) > & \overset{l_2,\ l_3}{\cdots\nearrow} & \Big|{\scriptstyle d_i} & \overset{l_1}{\nearrow} & \Big|{\scriptstyle p} \\ \Big\downarrow{\scriptstyle \iota_2} & \overset{w_2}{\searrow\nearrow} & \Big\downarrow & & \Big\downarrow \\ < (i+1,+), (i,+) > & \xrightarrow{w} & < (i,+) > & \longrightarrow & X \\ \Big\downarrow{\scriptstyle [s_{i+1}, s_i]} & & \Big\downarrow{\scriptstyle s_i} & & \\ \Delta^{n+1} & \xrightarrow{s_i} & \Delta^n & & \end{array}$$

Here we have pulled d_i back along s_i and then decomposed it vertically. This means that the map $w = \widehat{s_i}$ is the unique map making

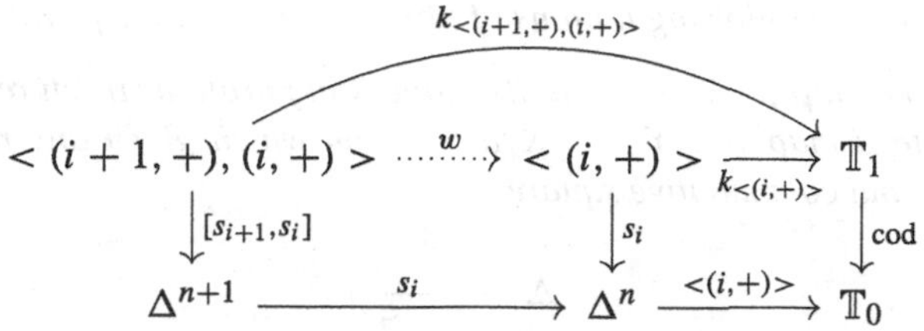

$$k_{<(i+1,+),(i,+)>} : \; <(i+1,+),(i,+)> \longrightarrow \mathbb{T}_1$$

$$\begin{array}{ccccc}
<(i+1,+),(i,+)> & \overset{w}{\dashrightarrow} & <(i,+)> & \xrightarrow[k_{<(i,+)>}]{} & \mathbb{T}_1 \\
\downarrow{\scriptstyle [s_{i+1}, s_i]} & & \downarrow{\scriptstyle s_i} & & \downarrow{\scriptstyle \mathrm{cod}} \\
\Delta^{n+1} & \xrightarrow{s_i} & \Delta^n & \xrightarrow{<(i,+)>} & \mathbb{T}_0
\end{array}$$

commute. Here $k_{<(i+1,+),(i,+)>} = [z_1, z_2]$ with z_1 choosing the position between the first and second element in $<(i+2,+),(i+1,+),(i,+)>$ and z_2 choosing the position between the second and third. Therefore $w = [s_i, s_{i+1}]$ and $w_2 = s_{i+1}$. This means that the top left square in the last but one diagram coincides with the left hand square in the diagram before that. Therefore also the lifts l_1 and l_2 in both diagram must coincide and since they are compatible with l_3, we deduce $l_1.s_{i+1} = l_2$, as desired. This completes the proof for the positive case: the negative case is similar. □

10.4 Naive Kan Fibrations in Simplicial Sets

From the previous theorem, we get two new descriptions of the naive Kan fibrations. Both start by observing that the entire lifting structure against $\mathbb{H}$ is already determined by a subclass of the vertical maps. First of all, we can consider those inclusions $<> \to \theta$ with empty domain: any other lift is completely determined by these, because

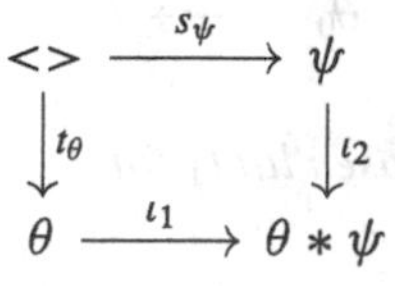

$$\begin{array}{ccc}
<> & \xrightarrow{s_\psi} & \psi \\
\downarrow{\scriptstyle \iota_\theta} & & \downarrow{\scriptstyle \iota_2} \\
\theta & \xrightarrow{\iota_1} & \theta * \psi
\end{array}$$

is a cocartesian square. Therefore the lift against the arrow on the right has to be the pushout of the lift against the arrow of the right. So one can equivalently define a naive Kan fibration structure in terms of lifts against arrows of the form $<> \to \theta$. If one does so, the horizontal compatibility condition for maps of the form $(1, \sigma)$ drops out and we are left with the horizontal compatibility condition for maps of the form $(\alpha, <>)$. In other words, we have:

Corollary 10.1 *The following notions of fibred structure are equivalent:*

- *To assign to a map $p : Y \to X$ all its naive Kan fibration structures.*
- *To assign to a map $p : Y \to X$ a function which given any n-dimensional traversal θ and commutative square*

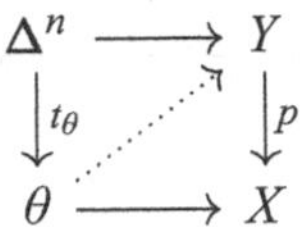

$$\begin{array}{ccc} \Delta^n & \longrightarrow & Y \\ \downarrow{\scriptstyle t_\theta} & \nearrow & \downarrow{\scriptstyle p} \\ \theta & \longrightarrow & X \end{array}$$

chooses a lift $\theta \to Y$. Moreover, these chosen lifts should satisfy two conditions:

(i) If $\alpha : \Delta^m \to \Delta^n$, then the chosen lifts

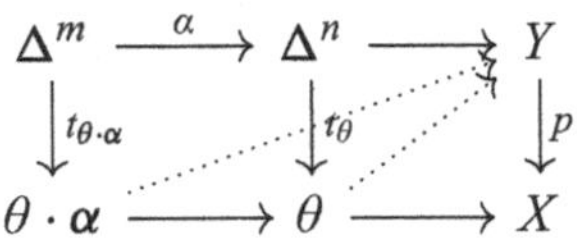

$$\begin{array}{ccccc} \Delta^m & \xrightarrow{\alpha} & \Delta^n & \longrightarrow & Y \\ \downarrow{\scriptstyle t_{\theta\cdot\alpha}} & & \downarrow{\scriptstyle t_\theta} & & \downarrow{\scriptstyle p} \\ \theta\cdot\alpha & \longrightarrow & \theta & \longrightarrow & X \end{array}$$

are compatible.

(ii) If $\theta = \theta_1 * \theta_0$, then the chosen lift $l : \theta \to Y$ can be computed in two steps: we can first compute the lift

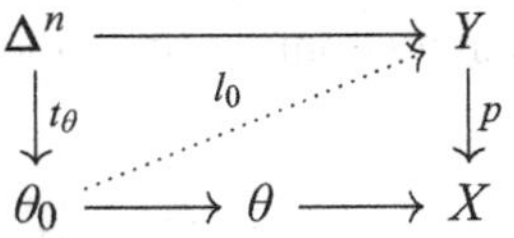

$$\begin{array}{ccccc} \Delta^n & & \longrightarrow & & Y \\ \downarrow{\scriptstyle t_\theta} & & l_0 & & \downarrow{\scriptstyle p} \\ \theta_0 & \longrightarrow & \theta & \longrightarrow & X \end{array}$$

and then compute the chosen lift l_1 for

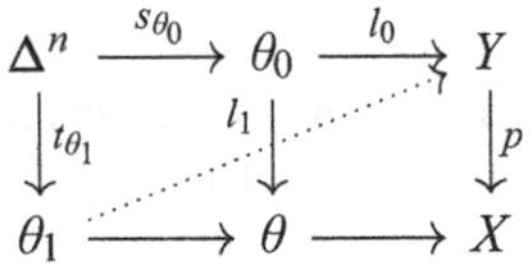

$$\begin{array}{ccccc} \Delta^n & \xrightarrow{s_{\theta_0}} & \theta_0 & \xrightarrow{l_0} & Y \\ \downarrow{\scriptstyle t_{\theta_1}} & & \downarrow{\scriptstyle l_1} & & \downarrow{\scriptstyle p} \\ \theta_1 & \longrightarrow & \theta & \longrightarrow & X \end{array}$$

and push this forward to obtain a map $l : \theta \to Y$ (so $l = [l_1, l_0]$).

In fact, if one sees the second notion of fibred structure can be seen as the vertical structure in a discretely fibred concrete double category in the obvious way, then this concrete double category is isomorphic to the concrete double category of naive Kan fibrations.

Using Corollary 9.2, the second bullet in the previous condition can be seen as an externalisation of conditions (1)–(3) in the definition of a naive fibration structure on p (see Definition 4.6). So the previous proposition says that there is a isomorphism of notions of fibred structure from maps carrying a naive fibration structure satisfying conditions (1)–(4) to those which only satisfy (1)–(3). It turns out that this isomorphism is just the forgetful map:

Corollary 10.2 *If $p : Y \to X$ is a map in simplicial sets, then any map*

$$L : Y \times_X MX \to MY$$

which satisfies conditions (1)–(3) for being a naive fibration structure on p as in Definition 4.6 also satisfies the remaining fourth condition.

Proof Let us call a map $L : Y \times_X MX \to MY$ a weak naive fibration structure if it only satisfies conditions (1)–(3). Then we know that there is an isomorphism between the transport structures on p and the weak naive fibration structures on p obtained by the operations studied in this chapter. Let us see how we get a transport structure from a weak naive fibration structure in this way. We start by extending the weak naive fibration structure to a right lifting structure against $\mathbb{H}$, which can then be extended to all HDRs. This can be used to find the transport structure t by solving the problem

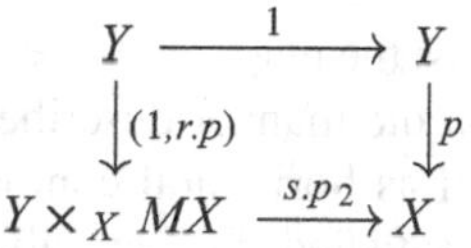

$$\begin{array}{ccc} Y & \xrightarrow{1} & Y \\ \downarrow{\scriptstyle (1,r.p)} & & \downarrow{\scriptstyle p} \\ Y \times_X MX & \xrightarrow{s.p_2} & X \end{array}$$

using that the map on the left is an HDR, via

$$\delta_p = (\alpha.(p_1, M!.p_2), \Gamma.p_2) : Y \times_X MX \to M(Y \times_X MX) \cong MY \times_{MX} MMX.$$

This means that on an arbitrary element $(y, (\theta, \pi : \theta \to X)) : \Delta^n \to Y \times_X MX$, the value $t(y, (\theta, \pi))$ is the solution to a lifting problem:

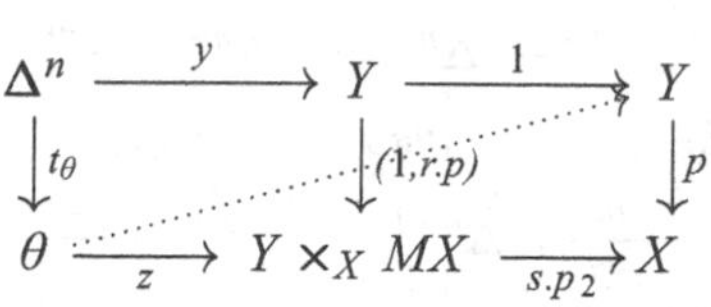

$$\begin{array}{ccccc} \Delta^n & \xrightarrow{y} & Y & \xrightarrow{1} & Y \\ \downarrow{\scriptstyle t_\theta} & & \downarrow{\scriptstyle (1,r.p)} & & \downarrow{\scriptstyle p} \\ \theta & \xrightarrow[z]{} & Y \times_X MX & \xrightarrow[s.p_2]{} & X \end{array}$$

where z is the transpose of $\delta_p(y, (\theta, \pi))$. But that means that $s.p_2.z$ is the transpose of $Ms.p_2.(\alpha.(p_1, M!.p_2), \Gamma.p_2)(y, (\theta, \pi)) = (\theta, \pi)$; in other words, $s.p_2.z = \pi$ and the induced lift is $L(y, (\theta, \pi))$. Therefore the induced transport structure is defined by $t(y, (\theta, \pi)) := s.L(y, (\theta, \pi))$.

So the upshot is that $L \mapsto s.L$ is the isomorphism of notion of fibred structure from the weak naive fibration structures to the transport structures. But we have seen in Proposition 4.10 that this also defines an isomorphism of notions of fibred structure between ordinary naive fibration structures and transport structures. We conclude that every weak naive fibration structure already satisfies condition (4). □

Remark 10.3 We believe that the previous corollary can also be shown directly. Very roughly, the reason is the following. One can think of Γ as being built from path composition and degeneracies, and since any weak naive fibration structure L is in particular a morphism of simplicial sets, it will automatically respect degeneracies. So if L respects path composition, it must also respect Γ.

If p is a naive Kan fibration, its lifting structure against $\mathbb{H}$ is also completely determined by its lifts against the inclusions of traversals of the form $<>\to<(i,\pm)>$. Indeed, we already used and explained this in the proof of Theorem 10.2: any vertical map in $\mathbb{H}$ is a vertical composition of inclusions of traversals where the next traversal has length one more than the previous and each such inclusion is a pushout of one of the form $<>\to<(i,\pm)>$. In the remainder of this chapter we will determine which compatibility conditions the lifts against these maps have to satisfy in order to extend to a (unique) lifting structure against $\mathbb{H}$. This description will also allow us to prove that the notion of a being a naive Kan fibration is a local notion of fibred structure.

If we are given the lifts against the maps $<>\to<(i,\pm)>$ and we extend them to the entire double category $\mathbb{H}$ in the manner described above, then both the vertical compatibility condition as well as horizontal compatibility condition for maps of form $(1,\sigma)$ are automatically satisfied. So we only need to ensure the horizontal compatibility condition for maps of the form $(\alpha, <>)$. To ensure that, we only need to consider squares where the horizontal map α is either a face or degeneracy maps and the vertical maps on the right is one of the form $<>\to<(i,\pm)>$.

We obtain the following cases:

(i) For the face maps, we have compatibility conditions for the case $k < i$ (left) and the case $k > i$ (right):

$$
\begin{array}{ccc}
\Delta^{n-1} & \xrightarrow{d_k} & \Delta^n \\
\downarrow{\scriptstyle d_{(i-1,\pm)^t}} & & \downarrow{\scriptstyle d_{(i,\pm)^t}} \\
\Delta^n & \xrightarrow{d_k} & \Delta^{n+1} \\
\downarrow{\scriptstyle s_{i-1}} & & \downarrow{\scriptstyle s_i} \\
\Delta^{n-1} & \xrightarrow{d_k} & \Delta^n
\end{array}
\qquad\qquad
\begin{array}{ccc}
\Delta^{n-1} & \xrightarrow{d_k} & \Delta^n \\
\downarrow{\scriptstyle d_{(i,\pm)^t}} & & \downarrow{\scriptstyle d_{(i,\pm)^t}} \\
\Delta^n & \xrightarrow{d_{k+1}} & \Delta^{n+1} \\
\downarrow{\scriptstyle s_i} & & \downarrow{\scriptstyle s_i} \\
\Delta^{n-1} & \xrightarrow{d_k} & \Delta^n
\end{array}
$$

What we mean here is that we have a horizontal compatibility condition for the top squares in both diagrams below, in that if $p : Y \to X$ is a naive Kan fibration, and we have lifting problem as in

$$\begin{array}{ccccc} \Delta^{n-1} & \xrightarrow{d_k} & \Delta^n & \longrightarrow & Y \\ \downarrow{\scriptstyle d_{(i-1,\pm)^t}} & & \downarrow{\scriptstyle d_{(i,\pm)^t}} & \nearrow & \downarrow{\scriptstyle p} \\ \Delta^n & \xrightarrow{d_k} & \Delta^{n+1} & \longrightarrow & X \\ \downarrow{\scriptstyle s_{i-1}} & & \downarrow{\scriptstyle s_i} & & \\ \Delta^{n-1} & \xrightarrow{d_k} & \Delta^n & & \end{array} \qquad \begin{array}{ccccc} \Delta^{n-1} & \xrightarrow{d_k} & \Delta^n & \longrightarrow & Y \\ \downarrow{\scriptstyle d_{(i,\pm)^t}} & & \downarrow{\scriptstyle d_{(i,\pm)^t}} & \nearrow & \downarrow{\scriptstyle p} \\ \Delta^n & \xrightarrow{d_{k+1}} & \Delta^{n+1} & \longrightarrow & X \\ \downarrow{\scriptstyle s_i} & & \downarrow{\scriptstyle s_i} & & \\ \Delta^{n-1} & \xrightarrow{d_k} & \Delta^n & & \end{array}$$

then the dotted lifts have to be compatible. (Note that there is also a case $k = i$, but it is trivially satisfied, because in that case we get the identity inclusion $<> \to <>$ on the left.)

(ii) For the degeneracy maps, we have compatibility condition for the case $k < i$ (left) and $k > i$ (right):

$$\begin{array}{ccc} \Delta^{n-1} & \xrightarrow{s_k} & \Delta^n \\ \downarrow{\scriptstyle d_{(i+1,\pm)^t}} & & \downarrow{\scriptstyle d_{(i,\pm)^t}} \\ \Delta^n & \xrightarrow{s_k} & \Delta^{n+1} \\ \downarrow{\scriptstyle s_{i+1}} & & \downarrow{\scriptstyle s_i} \\ \Delta^{n-1} & \xrightarrow{s_k} & \Delta^n \end{array} \qquad \begin{array}{ccc} \Delta^{n-1} & \xrightarrow{s_k} & \Delta^n \\ \downarrow{\scriptstyle d_{(i,\pm)^t}} & & \downarrow{\scriptstyle d_{(i,\pm)^t}} \\ \Delta^n & \xrightarrow{s_{k+1}} & \Delta^{n+1} \\ \downarrow{\scriptstyle s_i} & & \downarrow{\scriptstyle s_i} \\ \Delta^{n-1} & \xrightarrow{s_k} & \Delta^n \end{array}$$

as in (i).

(iii) Pulling back $< (i, \pm) >$ along s_i is a rather special case, which we split in both a positive and negative case (on the left and right, respectively).

$$\begin{array}{ccccc} & & \Delta^{n+1} & \xrightarrow{s_i} & \Delta^n \\ & & \downarrow{\scriptstyle d_i} & & \downarrow{\scriptstyle d_i} \\ \Delta^{n+1} & \xrightarrow{d_{i+1}} & \Delta^{n+2} & \searrow{\scriptstyle s_{i+1}} & \\ \downarrow{\scriptstyle d_{i+1}} & & \downarrow{\scriptstyle \iota_2} & & \\ \Delta^{n+2} & \xrightarrow{\iota_1} & \Delta^{n+2} \cup_{\Delta^{n+1}} \Delta^{n+2} & \xrightarrow{[s_i, s_{i+1}]} & \Delta^{n+1} \\ & & \downarrow{\scriptstyle [s_{i+1}, s_i]} & & \downarrow{\scriptstyle s_i} \\ & & \Delta^{n+1} & \xrightarrow{s_i} & \Delta^n \end{array} \qquad \begin{array}{ccccc} & & \Delta^{n+1} & \xrightarrow{s_i} & \Delta^n \\ & & \downarrow{\scriptstyle d_{i+2}} & & \downarrow{\scriptstyle d_{i+1}} \\ \Delta^{n+1} & \xrightarrow{d_{i+1}} & \Delta^{n+2} & \searrow{\scriptstyle s_i} & \\ \downarrow{\scriptstyle d_{i+1}} & & \downarrow{\scriptstyle \iota_2} & & \\ \Delta^{n+2} & \xrightarrow{\iota_1} & \Delta^{n+2} \cup_{\Delta^{n+1}} \Delta^{n+2} & \xrightarrow{[s_{i+1}, s_i]} & \Delta^{n+1} \\ & & \downarrow{\scriptstyle [s_i, s_{i+1}]} & & \downarrow{\scriptstyle s_i} \\ & & \Delta^{n+1} & \xrightarrow{s_i} & \Delta^n \end{array}$$

Therefore we obtain compatibility conditions as follows:

(a) In the positive case for:

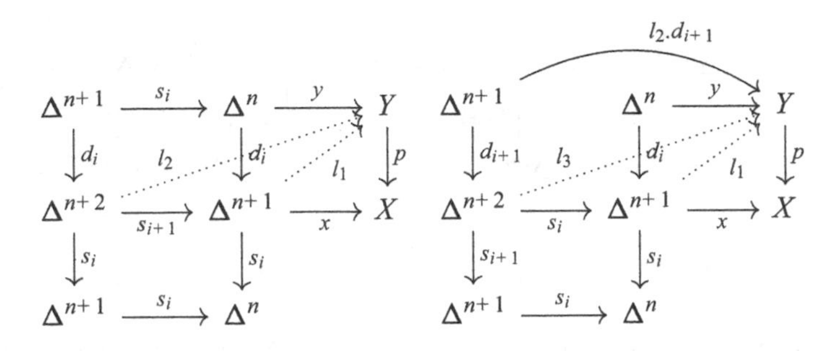

The diagram on the left expresses a compatibility condition similar to the previous ones (even in that the top left square is a morphism of HDRs: see the proof of Theorem 10.2). The one on the right is different, because there is no map $\Delta^{n+1} \to \Delta^n$ making the top left hand square commute. Note that the diagram on the left implies that $l_2.d_{i+1} = l_1.s_{i+1}.d_{i+1} = l_1$, so the reference to l_2 in the diagram on the right can be eliminated.

(b) In the negative case we have similar compatibility conditions:

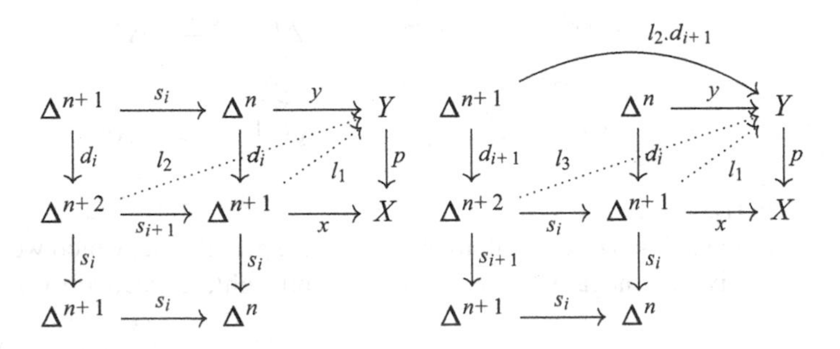

Remark 10.4 Note that in this notion of fibred structure we do not just choose lifts for each commutative square with some $d_i : \Delta^n \to \Delta^{n+1}$ on the right: we are also given as input a retraction of d_i (which has to be either s_{i-1} or $s_i : \Delta^{n+1} \to \Delta^n$). So although the lifting problem in no way refers to this retraction, the lifting structure may choose different solutions if d_i comes equipped with a different retraction. Also, the compatibility condition is formulated not for the d_i as such, but for the d_i together with a choice of retraction: indeed, the compatibility condition takes this choice into account in a crucial way.

From this characterisation we immediately get:

Corollary 10.3 *In the category of simplicial sets being a naive Kan fibration is a local notion of fibred structure.*

Another thing which this definition of a naive Kan fibration makes clear is that the traversals with positive and negative orientation live in parallel universes and there are no compatibility conditions relating the two. Indeed, to equip a map with the structure of a naive Kan fibration means equipping it with the structure of a naive right fibration and with the structure of a naive left fibration, with no requirements on how these two structures should relate. Put differently, we have:

Corollary 10.4 *In the category of notions of fibred structure, the notion of being a naive Kan fibration is the categorical product of the notion of being a naive right fibration and the notion of being a naive left fibration.*

Chapter 11
Mould Squares in Simplicial Sets

In this and the next chapter we will study *effective Kan fibrations* in simplicial sets. By definition, they are those maps which have the right lifting property against the large triple category of mould squares, with mould squares coming from the simplicial Moore path functor M and the pointwise decidable monomorphism as the cofibrations. The main aim of this chapter is to show that there is a small triple category of mould squares which generates the same class. In the next chapter we will use this to show that the effective Kan fibrations in simplicial sets form a local notion of fibred structure.

Remark 11.1 The attentive reader will notice that results similar to the ones we derive here hold for the mould squares coming from the two other Moore structures on simplicial sets (see Theorem 9.1 and Definition 9.2). We will refer to the maps having the right lifting property against the triple category of mould squares coming from (M_+, Γ_+, s) as the *effective right fibrations* and the maps having the right lifting property against the triple category of mould squares coming from (M_+, Γ_+^*, t) as the *effective left fibrations* . Implicitly we will show that these are also generated by suitable small triple categories of mould squares.

11.1 Small Mould Squares

We will define a triple category $\mathbb{M}$ as follows.

- Objects are triples (n, S, θ), usually just written (S, θ), consisting of a natural number n, a cofibrant sieve $S \subseteq \Delta^n$ and an n-dimensional traversal θ.
- There is a unique horizontal morphism $(S_0, \theta_0) \to (S_1, \theta_1)$ if $S_0 = S_1$ and θ_0 is a final segment of θ_1.

B. van den Berg, E. Faber, *Effective Kan Fibrations in Simplicial Sets*,
Lecture Notes in Mathematics 2321,
https://doi.org/10.1007/978-3-031-18900-5_11

- There is a unique vertical morphism $(S_0, \theta_0) \to (S_1, \theta_1)$ if $\theta_0 = \theta_1$ and $S_0 \subseteq S_1$ is an inclusion of cofibrant sieves.
- Perpendicular morphisms $(T \subseteq \Delta^m, \psi) \to (S \subseteq \Delta^n, \theta)$ are pairs (α, σ) with $\alpha : \Delta^m \to \Delta^n$ and σ an m-dimensional traversal such that $\alpha^* S = T$ and $\psi * \sigma = \theta \cdot \alpha$. Perpendicular composition is given by $(\alpha, \sigma).(\beta, \tau) = (\alpha.\beta, \tau * (\sigma \cdot \beta))$, as before.
- The triple category is codiscrete in the xy-plane in that whenever pairs of horizontal and vertical arrows fit together as in

$$\begin{array}{ccc} (S_0, \theta_0) & \longrightarrow & (S_0, \theta_1) \\ \downarrow & & \downarrow \\ (S_1, \theta_0) & \longrightarrow & (S_1, \theta_1), \end{array}$$

  then this is the boundary of a unique square. We will refer to such a square as a *small mould square*.
- In the yz- and xz-plane squares exist as soon as the perpendicular arrows have the same label (α, σ) (and the domains and codomains match up), and any two such which are "parallel" (have identical boundaries) are identical.
- The triple category is codiscrete in the third dimension, in that any potential boundary of a cube contains a unique cube filling it.

Proposition 11.1 *There is a triple functor* $\mathbb{M} \to \mathrm{MSq}(\widehat{\Delta})$ *from the triple category* $\mathbb{M}$ *to the large triple category of mould squares in simplicial sets.*

Proof Perhaps it is good to remind the reader of the structure of the large triple category of mould squares in simplicial sets:

- The objects are simplicial sets.
- The horizontal morphisms are HDRs.
- The vertical morphisms are cofibrations.
- The perpendicular morphisms are arbitrary maps of simplicial sets.
- The squares in the xy-plane are mould squares (morphisms of HDRs which are cartesian over a cofibration).
- The squares in the xz-plane are morphisms of HDRs.
- The squares in yz-plane are morphisms of cofibrations (that is, pullbacks).
- The cubes are pullback squares of HDRs (of a mould square along an arbitrary morphism of HDRs).

The idea is to send the object (n, S, θ) to the pullback $\theta \cdot S$ in simplicial sets:

$$\begin{array}{ccc} \theta \cdot S & \longrightarrow & \theta \\ \downarrow & & \downarrow{\scriptstyle j_\theta} \\ S & \longrightarrow & \Delta^n. \end{array}$$

In the x-direction we send $(S, \theta_0) \to (S, \theta_1)$ to the HDR we obtain by pullback:

$$\begin{array}{ccc} S & \longrightarrow & \Delta^n \\ \downarrow & & \downarrow \\ \theta_0 \cdot S & \longrightarrow & \theta_0 \\ \downarrow & & \downarrow \\ \theta_1 \cdot S & \longrightarrow & \theta_1. \end{array}$$

Note that both squares become cartesian squares of HDRs. Because pullback preserves composition of HDRs, this operation preserves composition in the x-direction. Similarly, in the y-direction we send $(S_0, \theta) \to (S_1, \theta)$ to the cofibration we obtain by pullback, as follows:

$$\begin{array}{ccccc} \theta \cdot S_0 & \longrightarrow & \theta \cdot S_1 & \longrightarrow & \theta \\ \downarrow & & \downarrow & & \downarrow \\ S_0 & \longrightarrow & S_1 & \longrightarrow & \Delta^n. \end{array}$$

From this it immediately follows that the squares in the xy-plane are sent to mould squares in simplicial sets.

The next step should be that squares on the left are sent to morphisms of HDRs and squares on the right to morphisms of cofibrations:

$$\begin{array}{ccc} (T, \psi) & \xrightarrow{(\alpha,\sigma)} & (S, \theta) \\ \downarrow & & \downarrow \\ (T, \psi') & \xrightarrow{(\alpha,\sigma)} & (S, \theta') \end{array} \qquad \begin{array}{ccc} (T, \psi) & \xrightarrow{(\alpha,\sigma)} & (S, \theta) \\ \downarrow & & \downarrow \\ (T', \psi) & \xrightarrow{(\alpha,\sigma)} & (S', \theta) \end{array}$$

We will again split this up in the case where $\alpha = 1$ and the case where $\sigma = <>$. If $\alpha = 1$, then $S = T$, $S' = T'$, $\theta = \psi * \sigma$ and $\psi' = \tau * \psi$ and $\theta' = \tau * \theta$ for some traversal τ. In this case, the square on the left is sent to pullback of the right hand square

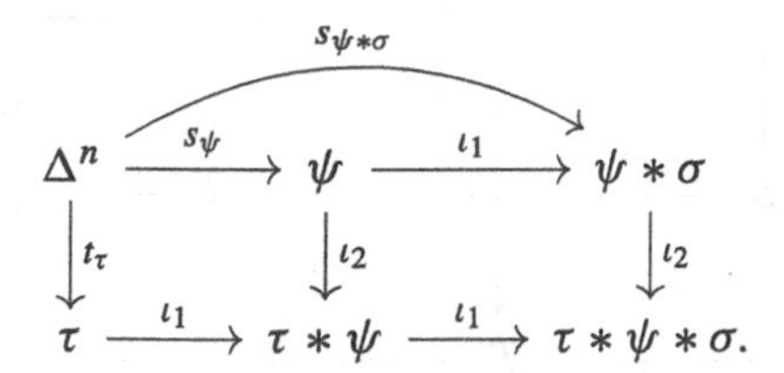

along $S \to \Delta^n$. Since pullback preserves bicartesian morphisms of HDRs (Beck-Chevalley!), the result is a bicartesian morphism of HDRs. In addition, since the outer rectangle and the right hand square in

$$
\begin{array}{ccccc}
\psi \cdot S & \longrightarrow & \theta \cdot S & \longrightarrow & S \\
\downarrow & & \downarrow & & \downarrow \\
\psi \cdot S' & \longrightarrow & \theta \cdot S' & \longrightarrow & S'
\end{array}
$$

are pullbacks, the square on the left is as well. Therefore the right hand square in the earlier diagram will be sent to a morphism of cofibrations when $\alpha = 1$.

Let us now consider the case $\sigma = <>$; now $\psi = \theta \cdot \alpha$ and $\psi' = \theta' \cdot \alpha$. Then we need to show that the front face of the bottom cube in

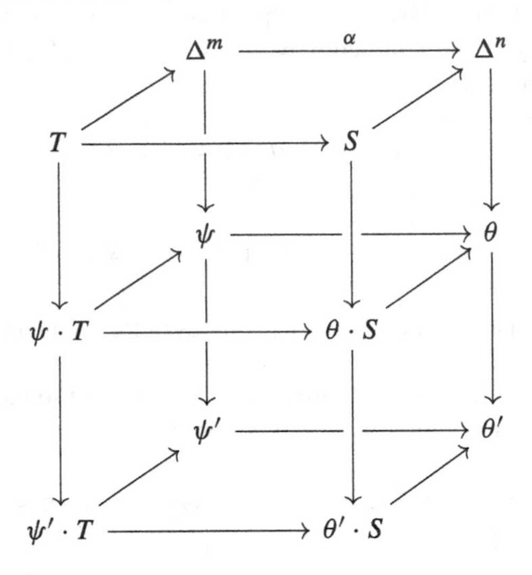

is a morphism of HDRs. But since all the other faces in this cube (besides the left and right one) are cartesian morphisms of HDRs, so must be the front face. Similarly, we need to show that the top of the left cube in

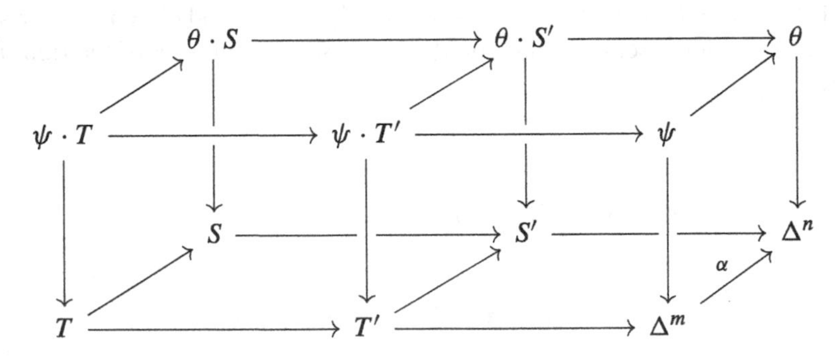

is a pullback. But it is not hard to see that all faces in both cubes must be pullbacks.

From the fact that the the squares in the yz-plane are pullbacks, it follows from Corollary 4.5 that the cubes are sent to pullback squares of HDRs. □

Remark 11.2 Note that it follows from the proof that the morphisms of HDRs that occur as images of squares in the xy-plane are Cartesian.

Theorem 11.1 *The following notions of fibred structure in simplicial sets are isomorphic:*

- *Having the right lifting property against the large triple category of mould squares (that is, to be an effective Kan fibration).*
- *To have the right lifting property against the small triple category* $\mathbb{M}$.

Indeed, the triple functor from $\mathbb{M}$ *to the large triple category of mould squares in simplicial sets induces a morphism of discretely fibred concrete double categories by right lifting properties: this induced morphism of concrete double categories satisfies fullness on squares and is therefore an isomorphism.*

Proof For reasons that will become clear later, we will first prove that both notions of fibred structure are equivalent if we ignore the vertical condition on both sides (so on both sides we have lifts satisfying only the horizontal and perpendicular conditions). After we have done that, we will show that the equivalence restricts to one where on both sides the vertical condition is satisfied as well.

So suppose $p : Y \to X$ has the right lifting property against the small mould squares satisfying the horizontal and perpendicular conditions, and assume we are given a lifting problem of the form

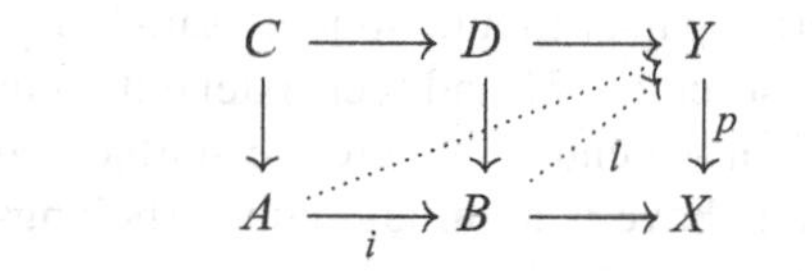

where the square on the left is a mould square. We wish to find a map $l : B \to Y$ making everything commute; for that, assume that we are given some $b \in B_n$. Let us write $(i : A \to B, j, H)$ for the HDR-structure on i. As we have seen in the previous chapter, we can construct a morphism of HDRs

$$\begin{array}{ccc} \Delta^n & \longrightarrow & A \\ \downarrow & & \downarrow{\scriptstyle i} \\ \theta & \xrightarrow{\pi} & B \end{array}$$

with $\pi.s_\theta = b$ and $j(b) = \theta$. By pulling back the mould square along this morphism of HDRs, we obtain a picture as follows:

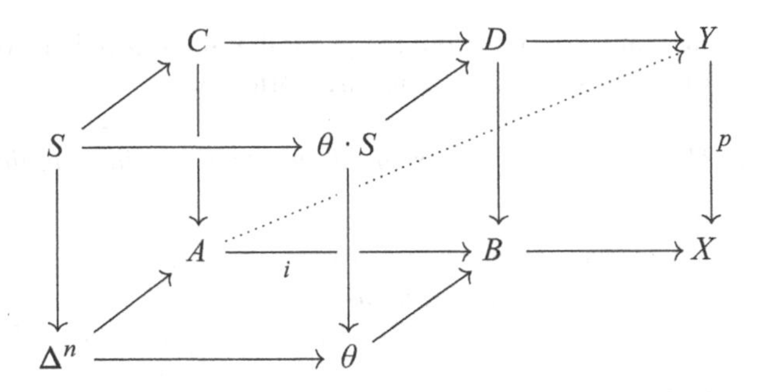

Since the mould square at the front of the cube belongs to $\mathbb{M}$, the picture induces a map $L_b : \theta \to Y$ making everything commute. We put $l(b) := L_b.s_\theta$, as in the previous chapter. At this point we need to verify a number of things: that this defines a natural transformation $B \to Y$, that this map fills the square and is compatible with the map $A \to Y$ that we were given. Also, we need to verify that if we choose these lifts for the mould squares, then together these lifts satisfy the horizontal and perpendicular compatibility conditions. Finally, we also need to verify fullness on squares. All of these things are just very minor extensions of results proved in the previous chapter, so we will omit the proofs here.

We will now show that the operation we have just defined and the one induced by the triple functor from the previous proposition are each other's inverses. One composite is clearly the identity: if we are given a map $p : Y \to X$ which has the right lifting property against mould squares satisfying the horizontal and perpendicular conditions, restrict it to $\mathbb{M}$ and then extend it all mould squares in the manner described above, then we end up where we started. The reason is simply that the cube in the diagram above is a "mould cube" (belongs to the large triple category of mould squares).

The converse is the hard bit: so imagine that we have a map $p : Y \to X$ which has the right lifting property against the small mould squares satisfying the horizontal and perpendicular conditions. This means that if we have a lifting problem of the form:

$$\begin{array}{ccccc} (S_0, \theta_0) & \longrightarrow & (S_0, \theta_1) & \longrightarrow & Y \\ \downarrow & & \downarrow & & \downarrow p \\ (S_1, \theta_0) & \longrightarrow & (S_1, \theta_1) & \longrightarrow & X, \end{array}$$

we can solve this problem in two different ways. First of all, we can use the lifting structure of p directly; but we can also observe that the square on the left is a large mould square and use the procedure outlined above to find the lift. The task is to show that both lifts are the same. Again, we argue as in the previous chapter, by first observing that we can reduce this problem to the situation where $\theta_0 =<>$ and $\theta_1 =< i, \pm >$. Indeed, we can write the mould square on the left as a horizontal composition of small mould squares where the traversal on the right has one entry more than the one on the left. Moreover, we have a mould cube

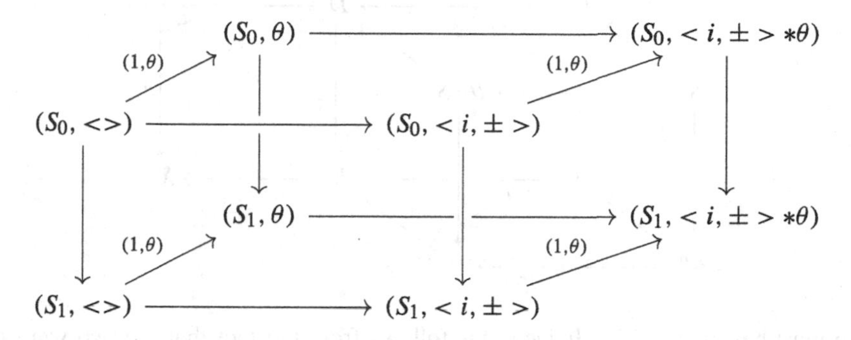

in which the top and bottom faces are cocartesian. Therefore the lift against the back is completely determined by that the lift against the front face. In fact, we can take it one step further: if $\alpha : \Delta^m \to \Delta^n \in S_1$, then

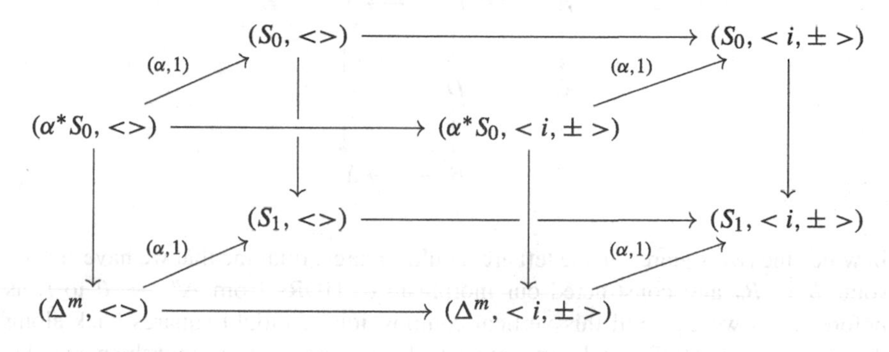

is a mould cube as well. This means that the lift against the back face is completely determined by the lifts against the front faces if we let α range over S_1. In short, we only have to compare lifts against small mould squares of the form:

$$\begin{array}{ccccc} (S, <>) & \longrightarrow & (S, < i, \pm >) & \longrightarrow & Y \\ \downarrow & & \downarrow & & \downarrow p \\ (\Delta^n, <>) & \longrightarrow & (\Delta^n, < i, \pm >) & \longrightarrow & X. \end{array}$$

But this can be argued for just as in the previous chapter, so we again omit the proof.

It remains to check that this equivalence of notions of fibred structure restricts to one where the vertical condition is satisfied one both sides. In fact, we only need to show that if $p : Y \to X$ comes equipped with lifts against the small mould squares (satisfying the vertical condition as well), and we extend this to all mould squares in the manner explained above, then the lifts against all the mould squares satisfy the vertical condition. Before we do that, we make the important point that in this extension of the lifting structure to all mould squares as in

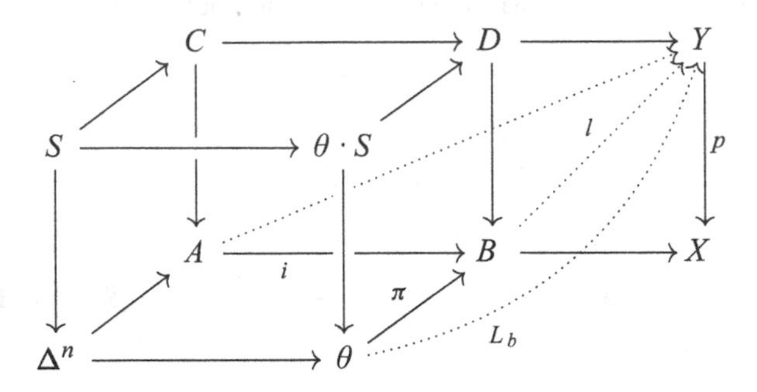

we must have $l.\pi = L_b$. Indeed, this follows from the fact that the two ways of computing of lifts against small mould squares coincide.

So imagine we have a lifting problem of the form:

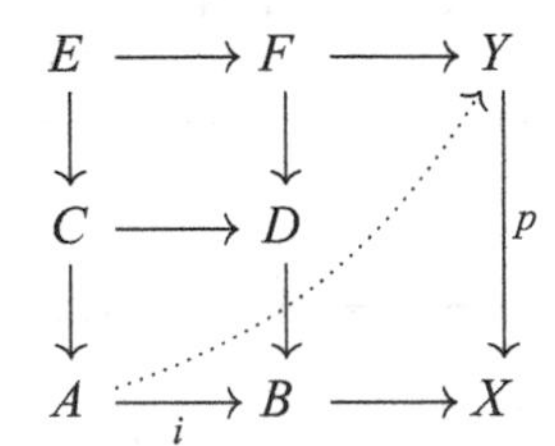

in which the two squares on the left are mould squares. Imagine that we have chosen some $b \in B_n$ and constructed our morphism of HDRs from $\Delta^n \to \theta$ to i, as before. Then we can pull this vertical composition of mould squares back along this morphism of HDRs, and pull that back along some arbitrary morphism $\alpha \in S_1$, as follows:

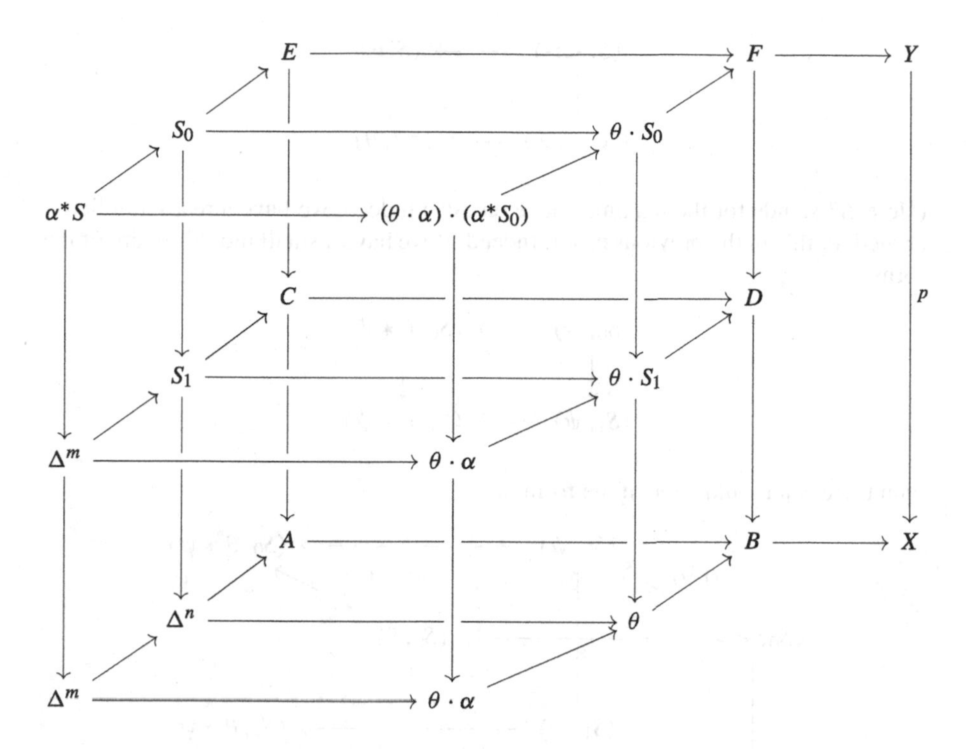

What this amounts to is saying that our lift $L_b : \theta \to Y$ can be computed by first computing the map $L_{b \cdot S_1} : \theta \cdot S_1 \to Y$. But that map is completely determined by the maps $L_{b \cdot \alpha} : \theta \cdot \alpha \to Y$ with α ranging over S_1. From this the vertical condition for the large mould squares at the back follows. □

11.2 Effective Kan Fibrations in Terms of "Filling"

If $p : Y \to X$ has the right lifting property against $\mathbb{M}$, then it comes equipped with a choice of lifts against every small mould square, where these lifts satisfy several compatibility conditions. Because of these compatibility conditions some of the lifts are completely determined by the choices we made for other lifts. What we can do is try to identify a suitable subclass and express the compatibility conditions purely in terms of lifts against elements in this smaller subclass. This is the game we have played already a number of times. For the small mould squares, we will take this to the limit in the next chapter, but here we can already note that the lifts general mould squares are completely determined by those of the form:

$$\begin{array}{ccc} (S, <>) & \longrightarrow & (S, \theta) \\ \downarrow & & \downarrow \\ (\Delta^n, <>) & \longrightarrow & (\Delta^n, \theta). \end{array}$$

(Here Δ^n stands for the maximal sieve on Δ^n.) Indeed, we have already implicitly argued for this in the previous proof. Indeed, if we have a small mould square of the form

$$\begin{array}{ccc} (S_0, \psi) & \longrightarrow & (S_0, \theta * \psi) \\ \downarrow & & \downarrow \\ (S_1, \psi) & \longrightarrow & (S_1, \theta * \psi) \end{array}$$

then there is a mould cube of the form

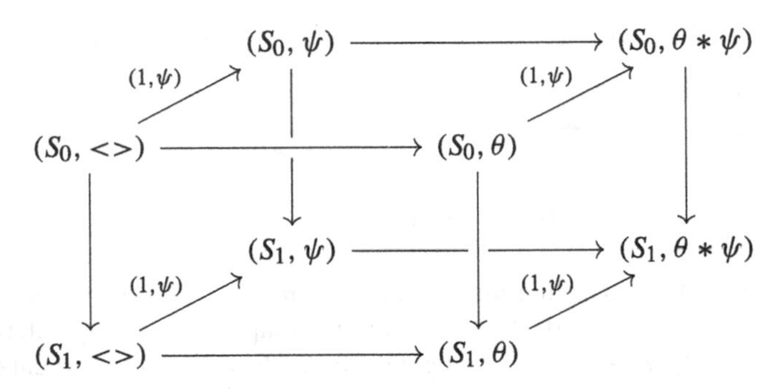

in which the top and bottom faces are cocartesian: therefore the lifts against the back in completely determined by the lift against the front. Furthermore, if we have a mould square as in the front of this mould cube, it occurs at the back of a mould cube

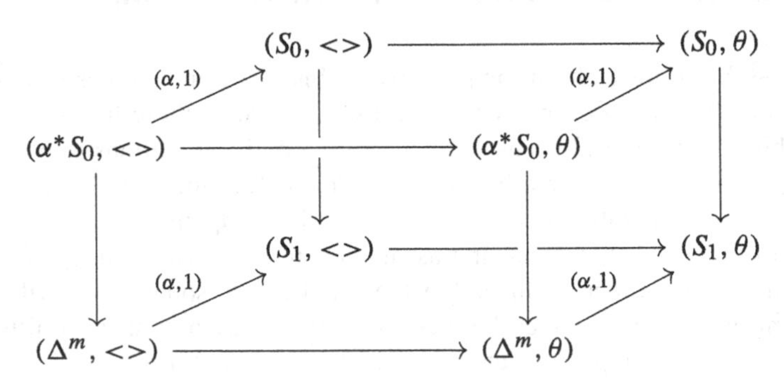

where $\alpha : \Delta^m \to \Delta^n \in S_1$. Since the $(\alpha, 1) : (\Delta^m, \theta) \to (S_1, \theta)$ collectively cover (S_1, θ), any compatible system of lifts against the front squares (while α ranges over S_1) descends to a unique lift against the front. Let us call the lift against the back that we obtain in this way *the induced lift*. Then we have the following result, whose proof we omit because it is a variation on a type of argument we have already seen a number of times.

Proposition 11.2 *The following notions of fibred structure are equivalent:*

- *To assign to each map $p : Y \to X$ all its effective Kan fibration structures.*
- *To assign to each map $p : Y \to X$ all functions which given a natural number $n \in \mathbb{N}$, a cofibrant sieve $S \subseteq \Delta^n$, an n-dimensional traversal θ and a commutative square*

$$\begin{array}{ccc} \Delta^n \cup \theta \cdot S & \xrightarrow{m} & Y \\ \downarrow{\scriptstyle [t_\theta, i_\theta]} & & \downarrow{\scriptstyle p} \\ \theta & \xrightarrow{n} & X \end{array}$$

choose a filler $\theta \to Y$. Moreover, these chosen fillers should satisfy the following three compatibility conditions:

(1) for each $\alpha : \Delta^m \to \Delta^n$ the choice of filler for the composed square

$$\begin{array}{ccccc} \Delta^m \cup (\theta \cdot \alpha) \cdot (\alpha^* S) & \longrightarrow & \Delta^n \cup \theta \cdot S & \xrightarrow{m} & Y \\ \downarrow & & \downarrow & & \downarrow{\scriptstyle p} \\ \theta \cdot \alpha & \longrightarrow & \theta & \xrightarrow{n} & X \end{array}$$

is the composition of $\theta \cdot \alpha \to \theta$ with the chosen filler $\theta \to Y$ for the right hand square.

(2) if $\theta = \theta_1 * \theta_0$, then the chosen filling for

$$\begin{array}{ccc} \Delta^n \cup \theta \cdot S & \xrightarrow{[y_0, [m_1, m_0]]} & Y \\ \downarrow & & \downarrow{\scriptstyle p} \\ \theta & \xrightarrow{n} & X \end{array}$$

coincides with the one we obtain in the following manner. One can first compute the filler for the composed square

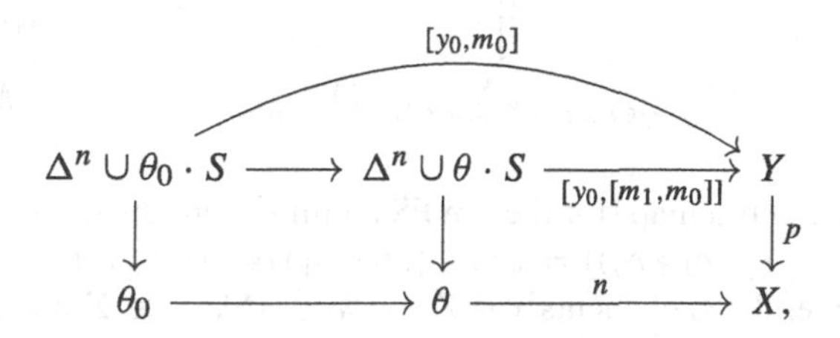

$$\begin{array}{ccccc} \Delta^n \cup \theta_0 \cdot S & \longrightarrow & \Delta^n \cup \theta \cdot S & \xrightarrow{[y_0, [m_1, m_0]]} & Y \\ \downarrow & & \downarrow & & \downarrow{\scriptstyle p} \\ \theta_0 & \longrightarrow & \theta & \xrightarrow{n} & X, \end{array}$$

(with a curved arrow $[y_0, m_0] : \Delta^n \cup \theta_0 \cdot S \to Y$ over the top) *from which we get an element* $y_1 : \Delta^n \to Y$ *by precomposition with the source map* $s_{\theta_0} : \Delta^n \to \theta_0$. *Then we can compute the filler for the square*

$$\begin{array}{ccc} \Delta^n \cup \theta_1 \cdot S & \xrightarrow{[y_1, m_1]} & Y \\ \downarrow & & \downarrow p \\ \theta_1 \longrightarrow \theta & \xrightarrow{n} & X. \end{array}$$

By amalgamating the two maps $\theta_i \to Y$ *we just constructed, we obtain another map* $\theta \to Y$, *which is the one which should coincide with the filler for the original square.*

(3) *if* $S_0 \subseteq S_1 \subseteq \Delta^n$ *then the chosen filler for*

$$\begin{array}{ccc} \Delta^n \cup \theta \cdot S_0 & \longrightarrow & Y \\ a \downarrow & & \\ \Delta^n \cup \theta \cdot S_1 & & \downarrow p \\ b \downarrow & & \\ \theta & \longrightarrow & X \end{array}$$

coincides with the one we obtain by first taking the induced lift $\Delta^n \cup \theta \cdot S_1 \to Y$ *of* p *against* a *and then the chosen lift* $\theta \to Y$ *of* p *against* b.

Using the description of M as a polynomial functor (see Corollary 9.2), one can also express the second item in the previous corollary as follows: to equip a map $p : Y \to X$ with the structure of an effective Kan fibration means choosing a map

$$L : \sum_{(y,\theta) \in Y \times_X MX} \sum_{\sigma \in \Sigma} MY^{\sigma}_{(y,\theta)} \to MY$$

such that:

1. L exhibits (t, Mp) as an effective trivial fibration, that is, L fills

$$\begin{array}{ccc} MY & \xrightarrow{1} & MY \\ \downarrow & \nearrow_{L} & \downarrow (t, Mp) \\ \sum_{(y,\theta) \in Y \times_X MX} \sum_{\sigma \in \Sigma} MY^{\sigma}_{(y,\theta)} & \longrightarrow & Y \times_X MX \end{array}$$

   and is an algebra map (for the AWFS coming from the dominance).
2. $L(y, \theta_1 * \theta_0, (\sigma, \rho_1 * \rho_0)) = L(y, \theta_1, (\sigma, \rho_1)) * L(s.L(y, \theta_1, (\sigma, \rho_1)), \theta_0, (\sigma, \rho_0))$ for all generalised elements $y \in Y$, $\theta_1, \theta_0 \in MX$, $\sigma \in \Sigma$ and $\rho_1, \rho_0 \in MY^{\sigma}$.

   From this we immediately obtain:

Corollary 11.1 *If $p : Y \to X$ is an effective Kan fibration, then*

$$(t, Mp) : MY \to Y \times_X MX$$

is an effective trivial fibration.

The fact that there are cartesian natural transformations $\iota^+, \iota^- : X^{\mathbb{I}} \to MX$ as in the previous chapter, means that we have pullback squares of the form

$$\begin{array}{ccc} Y^{\mathbb{I}} & \longrightarrow & MY \\ \big\downarrow{\scriptstyle (s/t, p^{\mathbb{I}})} & & \big\downarrow{\scriptstyle (t, Mp)} \\ Y \times_X X^{\mathbb{I}} & \longrightarrow & Y \times_X MX. \end{array}$$

And since effective trivial fibrations are stable under pullback, we can deduce:

Corollary 11.2 *If $p : Y \to X$ is an effective Kan fibration, then*

$$(s/t, p^{\mathbb{I}}) : Y^{\mathbb{I}} \to Y \times_X X^{\mathbb{I}}$$

are effective trivial fibrations. Therefore effective Kan fibrations are uniform Kan fibrations in the sense of [8].

In view of this result the comment in Box 1.1 should now make sense.

Chapter 12
Horn Squares

The purpose of this chapter is to show that our notion of an effective Kan fibration in simplicial sets is both local and classically correct. By the latter we mean that, in a classical metatheory, a map can be equipped with the structure of an effective Kan fibration precisely when it has the right lifting property against horn inclusions (the traditional notion of a Kan fibration). To prove both these statements we will use a characterisation of the effective Kan fibrations in terms of what we will call horn squares.

12.1 Effective Kan Fibrations in Terms of Horn Squares

Recall that the small mould squares are the squares in the yz-plane in the triple category $\mathbb{M}$ (see previous chapter).

Definition 12.1 A small mould square will be called a *one-step mould square* if in the horizontal direction the length of the traversal increases by one and in the vertical direction the sieve increases by one m-simplex which was not yet present, but all whose faces were. Among these one-step squares are the *horn squares*

$$\begin{array}{ccc} (\partial\Delta^n, \langle\rangle) & \longrightarrow & (\partial\Delta^n, \langle(i, \pm)\rangle) \\ \downarrow & & \downarrow \\ (\Delta^n, \langle\rangle) & \longrightarrow & (\Delta^n, \langle(i, \pm)\rangle) \end{array}$$

which start from the empty traversal in the horizontal direction and end with the maximal sieve in the vertical direction.

The reason for the name horn square is the following: a lifting problem for p against a horn square

B. van den Berg, E. Faber, *Effective Kan Fibrations in Simplicial Sets*,
Lecture Notes in Mathematics 2321,
https://doi.org/10.1007/978-3-031-18900-5_12

$$\begin{array}{ccccc}
(\partial\Delta^n, \langle\rangle) & \longrightarrow & (\partial\Delta^n, \langle(i,\pm)\rangle) & \longrightarrow & Y \\
\downarrow & & \downarrow & & \downarrow p \\
(\Delta^n, \langle\rangle) & \longrightarrow & (\Delta^n, \langle(i,\pm)\rangle) & \longrightarrow & X
\end{array}$$

is equivalent to a lifting problem for p against the map from the inscribed pushout of the left hand square to its bottom right corner:

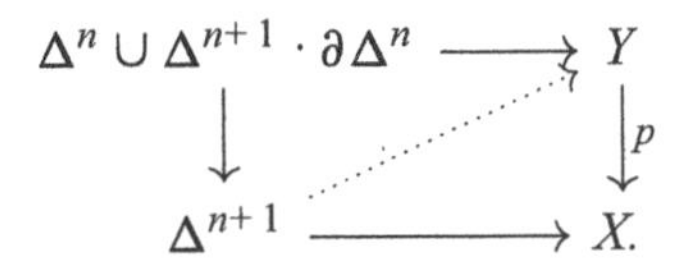

Here $\Delta^{n+1} \cdot \partial\Delta^n = \Lambda^{n+1}_{i,i+1} \cup (d_i \cap d_{i+1})$, that is, Δ^{n+1} with the interior and the ith and $(i+1)$st faces missing. Therefore in the previous square the map on left is the horn inclusion $\Lambda^{n+1}_{i+1} \to \Delta^{n+1}$ in the positive case and the horn inclusion $\Lambda^{n+1}_{i} \to \Delta^{n+1}$ in the negative case. Note that it follows from this that effective Kan fibration have the right lifting property against horn inclusions, so are Kan fibrations in the usual sense.

Remark 12.1 To make notation less cluttered, we will, from now on, write the traversal $< (i, \pm) >$ as $< i, \pm >$.

Lemma 12.1 *If a map is an effective Kan fibration in that it has the right lifting property against the triple category* $\mathbb{M}$ *of small mould squares, then this structure is completely determined by its lifts against the horn squares.*

Proof The proof combines three reductions, each of which we have already seen before.

First of all, any inclusion $S \subseteq T$ of sieves can be written as a sequence $S = S_0 \subseteq S_1 \subseteq S_2 \subseteq \ldots \subseteq S_n = T$ where at each point S_{i+1} is obtained from S_i by adding one new m-simplex whose boundary was already present in S_i. Therefore any small mould square can be decomposed into a grid

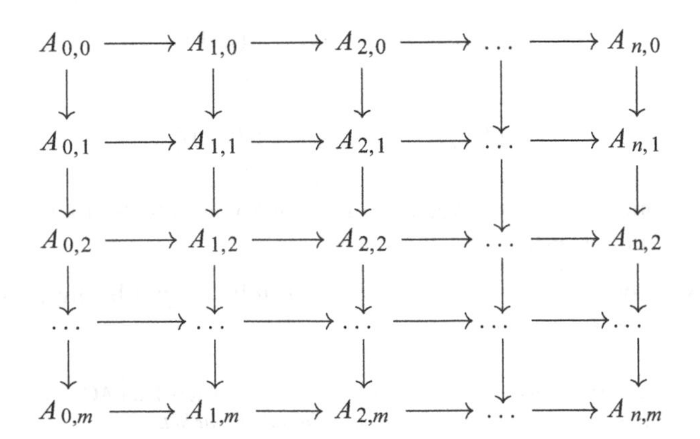

of one-step mould squares. Therefore the lifts against the one-step mould squares determine everything.

Secondly, as we have already seen in the proof of Theorem 11.1 the lifts against the one-step mould squares are determined by the one-step mould squares starting from the empty traversal. The reason, once again, is that there is a mould cube

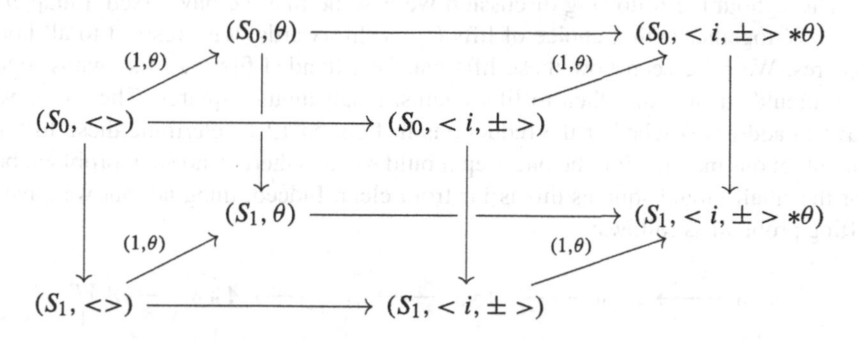

whose bottom face is a pushout.

Thirdly, suppose we have a one-step mould square starting from an empty traversal in the horizontal direction and suppose that in the vertical direction we have the inclusion $S \subseteq T$, where $\alpha : \Delta^m \to \Delta^n$ is the m-simplex that has been added to S to obtain T. Then

$$\begin{array}{ccc} \partial\Delta^m & \longrightarrow & S \\ \downarrow & & \downarrow \\ \Delta^m & \xrightarrow{\alpha} & T \end{array}$$

is bicartesian and we get a small mould cube of the form:

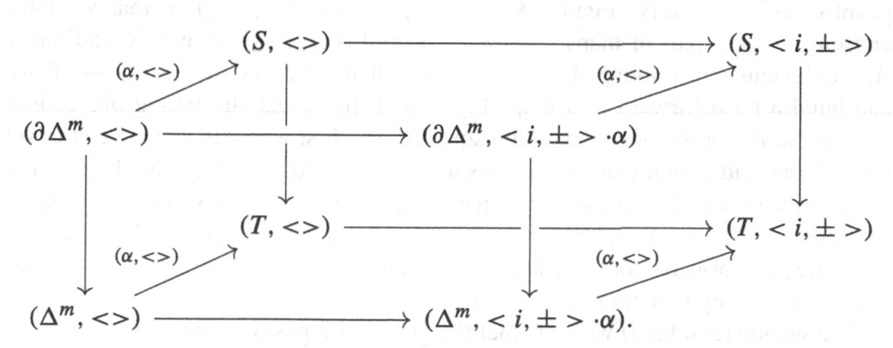

Since in this cube the right hand face is a pushout, the lift against the back face is determined by its front face. But because α is monic, the front face is either a horn square or trivial in the horizontal direction, depending on whether i is in the image of α or not. □

The remainder of this section will almost exclusively be devoted to answering the following question: suppose we are given a map p together with chosen lifts against the horn squares. Which conditions do these lifts have to satisfy in order for them to extend to a (necessarily unique) effective Kan fibration structure on p? We will answer this question in Theorem 12.1.

Throughout the following discussion we assume that we have fixed a map $p : Y \to X$ together with a choice of lifts (or pushforwards) with respect to all horn squares. We have seen how these lifts can be extended first to lifts against one-step mould squares and then to lifts against small mould squares. The worry we have to address is whether the reductions in Lemma 12.1 determine these lifts in unambiguous manner. For the one-step mould squares there is no such problem, but for the small mould squares this is far from clear. Indeed, imagine that we have a lifting problem as follows:

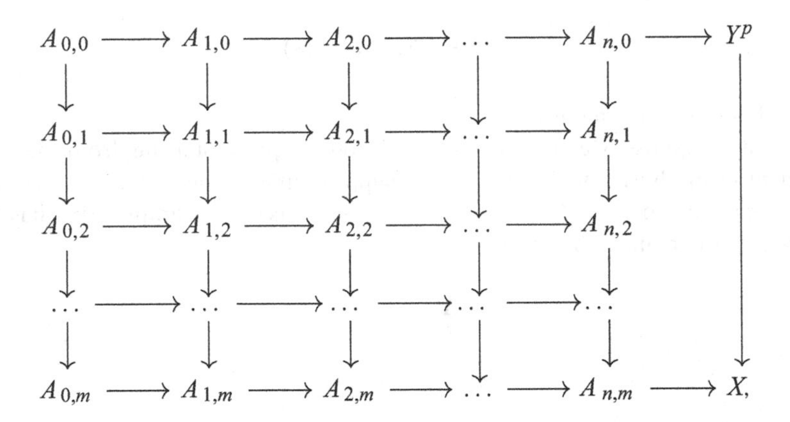

in which all the little squares are one-step mould squares. We have unambiguous pushforwards for every square $(A_{i,j}, A_{i+1,j}, A_{j+1,i}, A_{i+1,j+1})$ in that we have chosen for every pair of maps $A_{i+1,j} \to Y$ and $A_{i,j+1} \to Y$ over X and under $A_{i,j}$ an extension to a map $A_{i+1,j+1} \to Y$. Then for every map $A_{0,m} \to Y$ we can build a push forward to a map $A_{n,m} \to Y$ by repeatedly taking our chosen push forwards for the one-step mould squares. The first worry is that we can travel through the grid in many different ways and that it is not immediately obvious that we will always end up with the same map $A_{n,m} \to Y$. Still, this is the case, because if both $f_{i,j} : A_{i,j} \to Y$ and $g_{i,j} : A_{i,j} \to Y$ are obtained by repeatedly taking our favourite pushforwards for these little squares, in some order, then one easily proves that $f_{i,j} = g_{i,j}$ by induction on $n = i + j$.

The second (and final) worry is that the grid decomposing a small mould square into one-step mould squares is not uniquely determined. Clearly, we have no choice in how to travel in the horizontal direction, but in the vertical direction we have some choice, coming from the following fact. If $S \subseteq T \subseteq \Delta^n$ is an inclusion of cofibrant sieves and we write it as a sequence $S = S_0 \subseteq S_1 \subseteq S_2 \subseteq \ldots \subseteq S_n = T$ of cofibrant sieves where each S_{i+1} is obtained from S_i by adding a single m-simplex whose faces belonged to S_i, then this sequence is far from unique. However, any two such sequences can be obtained from each other by repeatedly applying permutations of the following form: if we have such a sequence and somewhere in this sequence we have $U \subseteq V \subseteq W$ where V is obtained from U by adding some k-simplex and W is obtained from W by adding some l-simplex and the k-simplex is not a face of the l-simplex (so that the boundaries of both the l-simplex and k-simplex were already present in U), then we can replace this by $U \subseteq V' \subseteq W$ where V' is obtained from U by adding the l-simplex and W is obtained from V' by adding the k-simplex. Since our answer to the first worry tells us that we may always assume that the way one finds the lift against a grid as above is by computing all the lifts $A_{i,j} \to Y$ in lexicographic order, we end up with the following statement that we need to prove:

Lemma 12.2 *Suppose U, V, V', W are cofibrant sieves as above and we have a lifting problem of the form:*

$$\begin{array}{ccccc}
(U,\theta) & \longrightarrow & (U, < i, \pm > *\theta) & \longrightarrow & Y \\
\downarrow & & \downarrow & & \downarrow p \\
(W,\theta) & \longrightarrow & (W, < i, \pm > *\theta) & \longrightarrow & X
\end{array}$$

then the solutions obtained by decomposing the left hand square as in the diagram below on the left or as in the one below on the right coincide.

$$\begin{array}{ccccc}
(U,\theta) & \longrightarrow & (U, < i, \pm > *\theta) & \longrightarrow & Y \\
\downarrow & & \downarrow & & \\
(V,\theta) & \longrightarrow & (V, < i, \pm > *\theta) & & \downarrow p \\
\downarrow & & \downarrow & & \\
(W,\theta) & \longrightarrow & (W, < i, \pm > *\theta) & \longrightarrow & X
\end{array}
\qquad
\begin{array}{ccccc}
(U,\theta) & \longrightarrow & (U, < i, \pm > *\theta) & \longrightarrow & Y \\
\downarrow & & \downarrow & & \\
(V',\theta) & \longrightarrow & (V', < i, \pm > *\theta) & & \downarrow p \\
\downarrow & & \downarrow & & \\
(W,\theta) & \longrightarrow & (W, < i, \pm > *\theta) & \longrightarrow & X
\end{array}$$

Proof Without loss of generality we may assume that θ is the empty traversal, the reason being that the solutions for general θ are obtained by pushing forward the solutions for $\theta =<>$.

Let us write $\alpha : [k] \to [n]$ for the k-simplex in V but not in U and $\beta : [l] \to [n]$ for the l-simplex in W but not in V. Then the solutions for the lifting problem in the left hand squares in both diagrams above are determined by the following mould cubes:

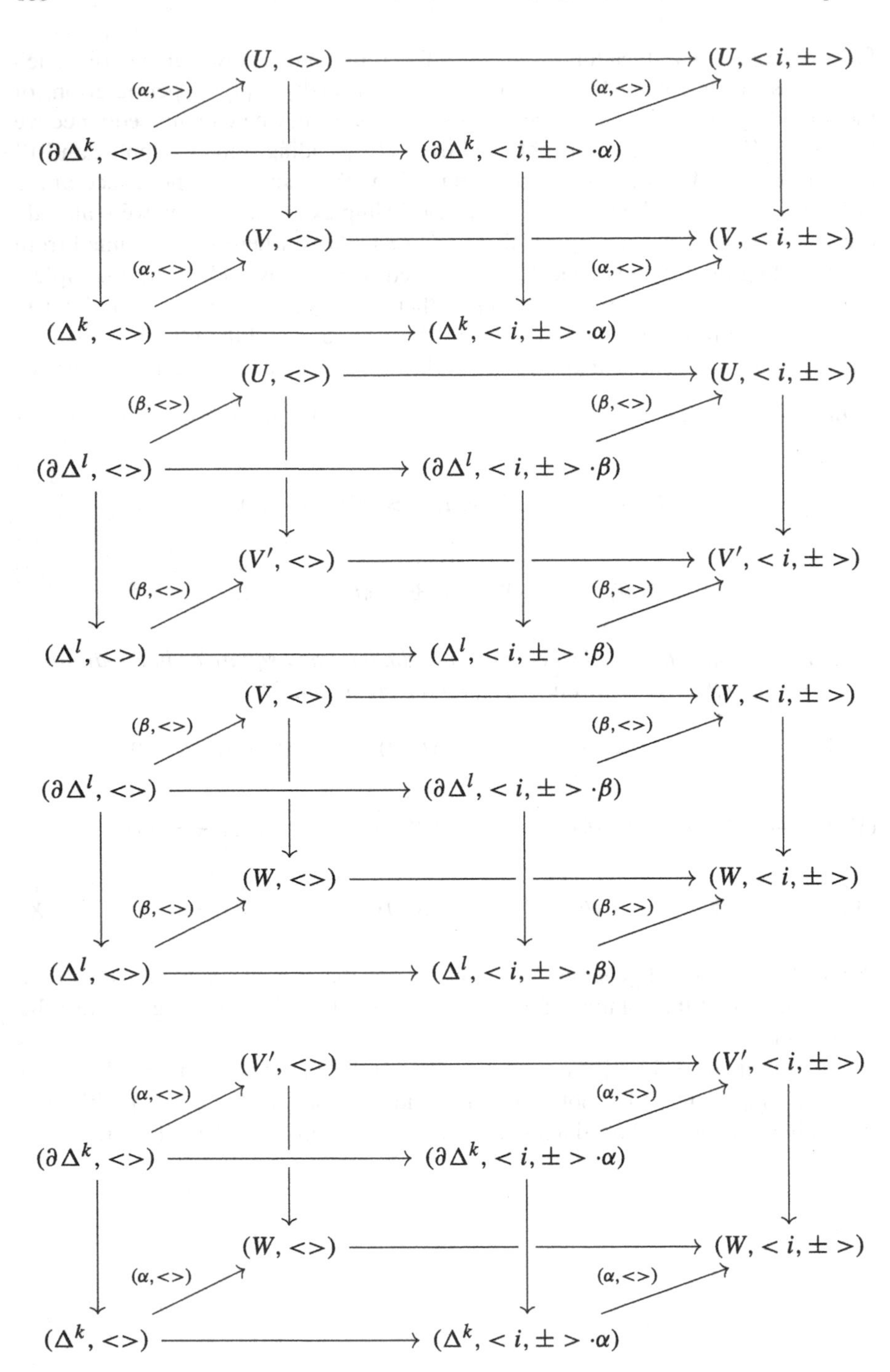

$(U, <>)$
$(U, < i, \pm >)$
$(\alpha, <>)$
$(\alpha, <>)$
$(\partial\Delta^k, <>)$
$(\partial\Delta^k, < i, \pm > \cdot\alpha)$
$(V, <>)$
$(V, < i, \pm >)$
$(\alpha, <>)$
$(\alpha, <>)$
$(\Delta^k, <>)$
$(\Delta^k, < i, \pm > \cdot\alpha)$
$(U, <>)$
$(U, < i, \pm >)$
$(\beta, <>)$
$(\beta, <>)$
$(\partial\Delta^l, <>)$
$(\partial\Delta^l, < i, \pm > \cdot\beta)$
$(V', <>)$
$(V', < i, \pm >)$
$(\beta, <>)$
$(\beta, <>)$
$(\Delta^l, <>)$
$(\Delta^l, < i, \pm > \cdot\beta)$
$(V, <>)$
$(V, < i, \pm >)$
$(\beta, <>)$
$(\beta, <>)$
$(\partial\Delta^l, <>)$
$(\partial\Delta^l, < i, \pm > \cdot\beta)$
$(W, <>)$
$(W, < i, \pm >)$
$(\beta, <>)$
$(\beta, <>)$
$(\Delta^l, <>)$
$(\Delta^l, < i, \pm > \cdot\beta)$
$(V', <>)$
$(V', < i, \pm >)$
$(\alpha, <>)$
$(\alpha, <>)$
$(\partial\Delta^k, <>)$
$(\partial\Delta^k, < i, \pm > \cdot\alpha)$
$(W, <>)$
$(W, < i, \pm >)$
$(\alpha, <>)$
$(\alpha, <>)$
$(\Delta^k, <>)$
$(\Delta^k, < i, \pm > \cdot\alpha)$

But since the second and third as well as the first and fourth have the same front face, the solutions will coincide. (What this says is that because the k-simplex and l-simplex have at most parts of their boundary in common, the the lifting problems are independent from each other and can be solved in either order.) □

To summarise the discussion so far, any map p which has the right lifting property with respect to horn squares, has unambiguous lifts against general small mould squares. Note that these lifts will automatically satisfy the horizontal and vertical conditions for having the right lifting property against the triple category $\mathbb{M}$, because the lifts do not depend on the way we divide a small mould square into a grid of one-step mould squares or on the way we traverse that grid. That means that any requirements on the lifts against the horn squares needed for them to extend to a unique effective Kan fibration structure should come from the perpendicular condition. So we will now have a look at this condition.

Remark 12.2 From now on, we will often think of the perpendicular condition on the lifts against the small mould squares in $\mathbb{M}$ as expressing as a stability condition with respect to base change or pullback. Indeed, we will often refer to it as a *base change condition*. The reason is that in $\mathbb{M}$ both the morphisms of HDRs in the yz-plane as well as in the xz-plane are cartesian. This means that in a small mould cube like

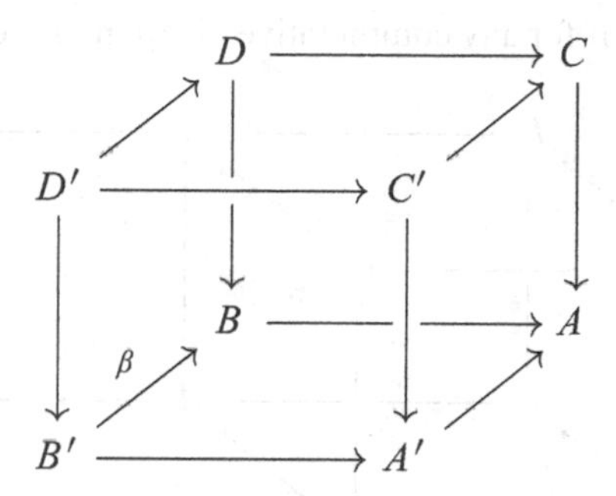

we can think of the front face of the cube as the result of pulling back the face at the back along $\beta : B' \to B$. Indeed, from now on we will often draw such a situation as follows

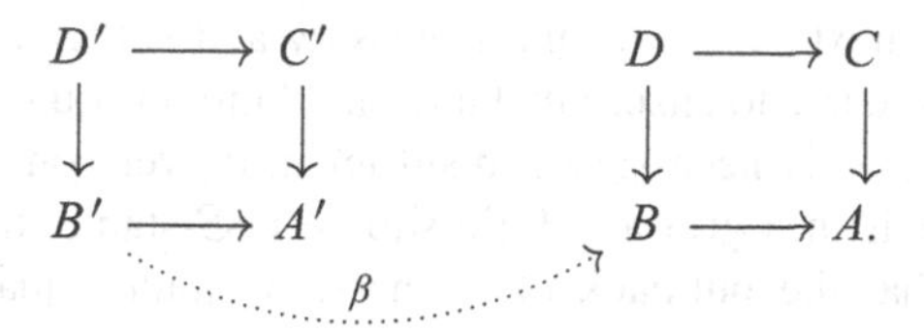

The reader is supposed to keep in mind that there is a cube connecting the two squares, but we will only draw a dotted arrow to prevent our diagrams from becoming too cluttered.

Let us call a small mould square *stable* if it its induced lift is compatible with the induced lift of any base change of that same square. In fact, it will be convenient to have a relativised notion of stability. So assume $\mathcal{S}$ is a class of morphisms in $\mathbf{\Delta}$ such that if

$$\begin{array}{ccc} [n'] & \xrightarrow{\alpha'} & [m'] \\ {\scriptstyle \beta'}\downarrow & & \downarrow{\scriptstyle \beta} \\ [n] & \xrightarrow[\alpha]{} & [m] \end{array}$$

is a pullback diagram in $\mathbf{\Delta}$ with β monic, then $\alpha \in \mathcal{S}$ implies $\alpha' \in \mathcal{S}$ (think $\mathcal{S} = \{$ face maps $\} \cup \{$ identities$\}$ or $\mathcal{S} = \{$ degeneracy maps $\} \cup \{$ identities$\}$). Then a small mould square

$$\begin{array}{ccc} D & \longrightarrow & C \\ \downarrow & & \downarrow \\ B & \longrightarrow & A \end{array}$$

will be called $\mathcal{S}$-*stable* if for any commutative diagram of the form

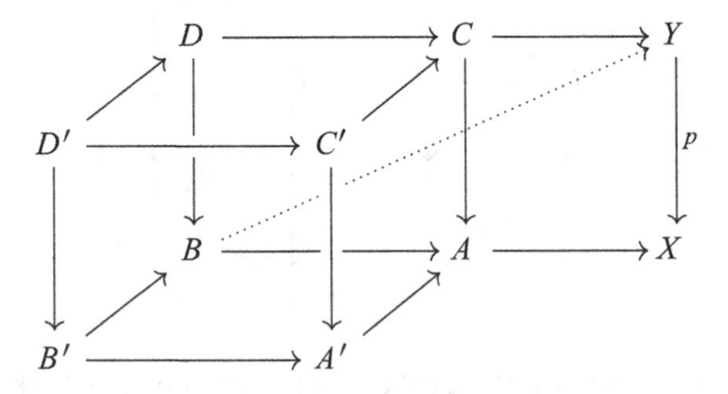

in which the cube is a "small" base change cube along $(\alpha, \tau) : B' \to B$ with $\alpha \in \mathcal{S}$, the induced map $A' \to Y$ can be obtained by composing the induced map $A \to Y$ with $A' \to A$. The next step is to find necessary and sufficient conditions on the fillers for the horn squares to ensure that any small mould square is $\mathcal{S}$-stable.

First of all, it is clearly necessary and sufficient if every one-step mould square is $\mathcal{S}$-stable. Indeed, if in a grid any little square is $\mathcal{S}$-stable, then so is the entire square. But note that the pullback of a one-step mould square need no longer be a one-step mould square: both in the horizontal and the vertical direction the number of steps may increase. (By the way, it may also become 0 in one of the two directions, in which case the stability condition is vacuously satisfied. So without loss of generality we may always assume this does not happen.)

Lemma 12.3 *Suppose $\psi = \tau * \theta$ and we have situation as follows:*

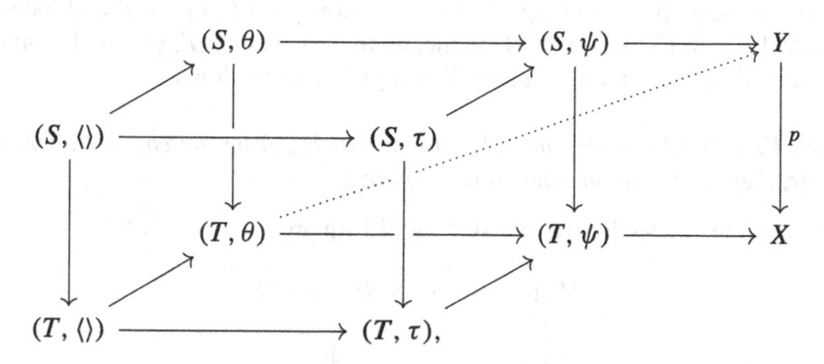

where the cube is a base change cube along $(1, \tau) : (T, \langle\rangle) \to (T, \theta)$. Then the induced lifts $(T, \tau) \to Y$ and $(T, \psi) \to Y$ are compatible. This means that, since the bottom of the cube is a pushout, the induced lifts determine each other by composition and pushout, respectively.

Proof We prove the statement of the lemma by induction on the length of the traversal τ. Note that the case $\tau = \langle\rangle$ is vacuously true.

In case where τ has length 1, we can regard both the front and the back of the cube as a vertical composition of one-step mould squares. In that case the statement follows from the definition of the induced lifts for one-step mould squares.

Now write $\tau = \sigma * \rho$ where σ has length 1 and consider the following situation:

$$\begin{array}{ccccccc}
 & & (S, \langle\rangle) & \longrightarrow & (S, \sigma) & & \\
 & & \downarrow & & \downarrow & & \\
 & & (T, \langle\rangle) & \longrightarrow & (T, \sigma) & & \\
 & & \vdots\ {\scriptstyle (1,\rho)} & & & & \\
(S, \langle\rangle) & \longrightarrow & (S, \rho) & \longrightarrow & (S, \tau) & & \\
\downarrow & & \downarrow & & \downarrow & & \\
(T, \langle\rangle) & \longrightarrow & (T, \rho) & \longrightarrow & (T, \tau) & & \\
\vdots\ {\scriptstyle (1,\theta)} & & \vdots\ {\scriptstyle (1,\theta)} & & & & \\
(S, \theta) & \longrightarrow & (S, \rho * \theta) & \longrightarrow & (S, \psi) & \longrightarrow & Y \\
\downarrow & & \downarrow & & \downarrow & & \downarrow p \\
(T, \theta) & \longrightarrow & (T, \rho * \theta) & \longrightarrow & (T, \psi) & \longrightarrow & X
\end{array}$$

We should imagine that we are given a map $(T, \theta) \to Y$ and we want to push it forward to a map $(T, \psi) \to Y$. Now, by induction hypothesis, the fact that the statement holds in case τ has length 1, and the earlier lemmas about grids, we can compute this as follows: take the induced lift $(T, \rho) \to Y$, push that down to map $(T, \rho*\theta) \to Y$, restrict that to a map $(T, \langle\rangle) \to Y$, take the induced lift $(T, \sigma) \to Y$

and then push that all the way down to a map $(T, \psi) \to Y$. But the latter can be done in two steps: push the map $(T, \sigma) \to Y$ down to $(T, \tau) \to Y$ and then push it further down to $(T, \psi) \to Y$. This means the end result $(T, \psi) \to Y$ coincides with taking the induced lift $(T, \tau) \to Y$ and pushing that down. □

Lemma 12.4 *If all one-step mould squares starting from the empty traversal are $\mathcal{S}$-stable, then so are all one-step mould squares.*

Proof Imagine that we have a one-step mould square

$$\begin{array}{ccc} (S, \theta) & \longrightarrow & (S, \langle i, \pm\rangle * \theta) \\ \downarrow & & \downarrow \\ (T, \theta) & \longrightarrow & (T, \langle i, \pm\rangle * \theta) \end{array}$$

and we pull it back along (α, σ) with $\alpha \in \mathcal{S}$. Writing $\theta \cdot \alpha = \psi * \sigma$, we get four small mould squares:

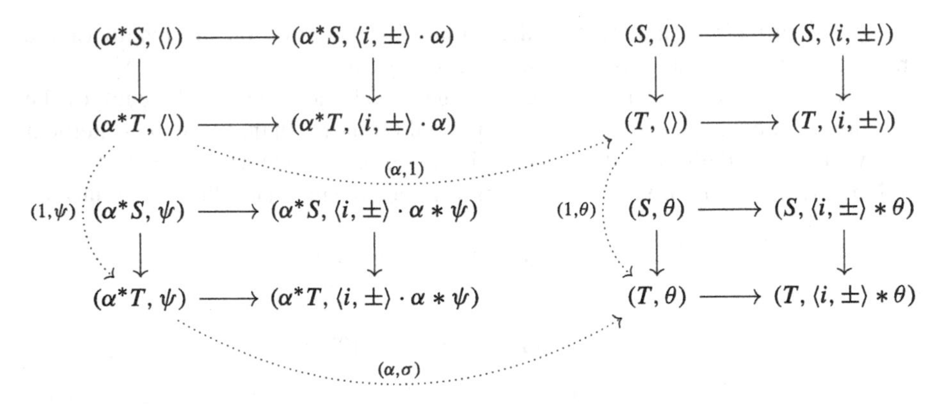

Recall from Remark 12.2 that the dotted arrows indicate that the squares are connected by small mould cubes and note that the small mould cubes determined by the dotted arrows going down have pushouts at their bottom faces. We are given a map $(T, \theta) \to Y$ and asked to compare the induced maps $(T, \langle i, \pm\rangle * \theta) \to Y$ and $(\alpha^* T, \langle i, \pm\rangle \cdot \alpha * \psi) \to Y$. The previous lemma tells us that both induced maps can be computed by taking the induced maps $(\alpha^* T, \langle i, \pm\rangle \cdot \alpha) \to Y$ and $(T, \langle i, \pm\rangle) \to Y$ and then pushing these down. So if the square on the top right is $\mathcal{S}$-stable, then so is the square on the bottom right. □

Lemma 12.5 *Suppose $S \subseteq T \subseteq \Delta^m$ are cofibrant sieves, $\alpha : \Delta^n \to \Delta^m$ is monic and*

$$\begin{array}{ccc} R & \longrightarrow & S \\ \downarrow & & \downarrow \\ \Delta^n & \xrightarrow{\alpha} & T \end{array}$$

is bicartesian. Suppose moreover that we have situation as follows:

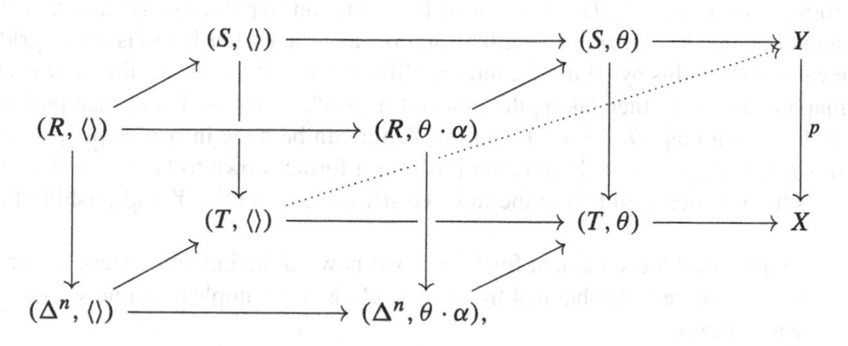

where the cube is a base change cube along $(\alpha, \langle\rangle) : (\Delta^n, \langle\rangle) \to (T, \langle\rangle)$*. Then the induced lifts* $(T, \theta) \to Y$ *and* $(\Delta^n, \theta \cdot \alpha) \to Y$ *are compatible. This means that, since the left and right hand faces of the cube are pushouts, the induced lifts determine each other by composition and pushout, respectively.*

Proof We prove this by induction on the number k of simplices in T but not in S (which coincides with the number of simplices in Δ^n but not in R). Note that the case $k = 0$ is trivial.

In case $k = 1$, we have $R = \partial\Delta^n$. We prove the desired statement by induction on the length of θ. Note that because α is monic, $\theta \cdot \alpha$ cannot have greater length than θ. The case $\theta = \langle\rangle$ is again trivial, while the case where θ has length 1 follows immediately from the way the lifts for horn squares induce lifts for one-step mould square starting from the empty traversal. Now write $\theta = \tau * \sigma$ where τ has length 1 and consider:

$$\begin{array}{ccc}
(R, \langle\rangle) & \longrightarrow & (R, \tau \cdot \alpha) \\
\downarrow & & \downarrow \\
(\Delta^n, \langle\rangle) & \longrightarrow & (\Delta^n, \tau \cdot \alpha)
\end{array}$$

$$(1, \sigma \cdot \alpha) : (\Delta^n, \tau \cdot \alpha) \dashrightarrow (\Delta^n, \sigma \cdot \alpha)$$

$$\begin{array}{ccccc}
(R, \langle\rangle) & \longrightarrow & (R, \sigma \cdot \alpha) & \longrightarrow & (R, \theta \cdot \alpha) \\
\downarrow & & \downarrow & & \downarrow \\
(\Delta^n, \langle\rangle) & \longrightarrow & (\Delta^n, \sigma \cdot \alpha) & \longrightarrow & (\Delta^n, \theta \cdot \alpha)
\end{array}$$

$$(\alpha, \langle\rangle) : (\Delta^n, \langle\rangle) \dashrightarrow (T, \langle\rangle)$$

$$\begin{array}{ccccccc}
(S, \langle\rangle) & \longrightarrow & (S, \sigma) & \longrightarrow & (S, \theta) & \longrightarrow & Y \\
\downarrow & & \downarrow & & \downarrow & & \downarrow p \\
(T, \langle\rangle) & \longrightarrow & (T, \sigma) & \longrightarrow & (T, \theta) & \longrightarrow & X
\end{array}$$

We should imagine that we are given a map $(T, \langle\rangle) \to Y$ and we want to push it forward to a map $(T, \theta) \to Y$. Now, by induction hypothesis, the fact that the statement holds in case θ has length 1 (and 0), and the earlier lemmas about grids, we can compute this by taking the induced lift $(\Delta^n, \sigma \cdot \alpha) \to Y$, pushing it down to a map $(T, \sigma) \to Y$, then taking the induced lift $(\Delta^n, \tau \cdot \alpha) \to Y$ and then pushing that down to a map $(T, \theta) \to Y$. But the latter can be done in two steps: pushing it down to $(\Delta^n, \theta \cdot \alpha) \to Y$ and then pushing it further down to $(T, \theta) \to Y$. This means it coincides with taking the induced lift $(\Delta^n, \theta \cdot \alpha) \to Y$ and pushing that down.

Having proved the statement for $k = 1$, we now do the induction step. So write $S \subseteq S' \subseteq T$ where S' is obtained from S by adding one simplex, so that we have a picture as follows:

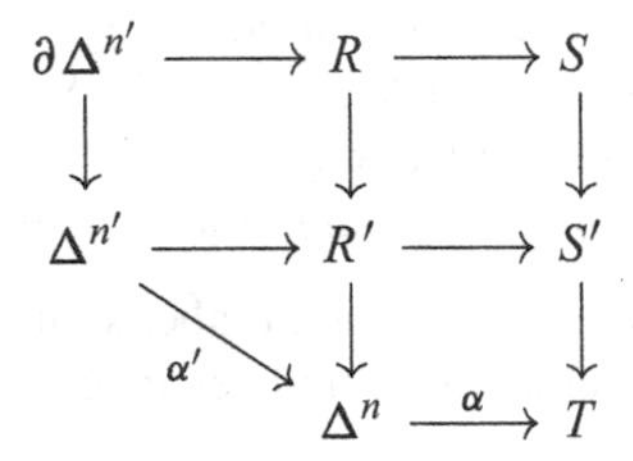

in which all squares are bicartesian. This gives us the following situation:

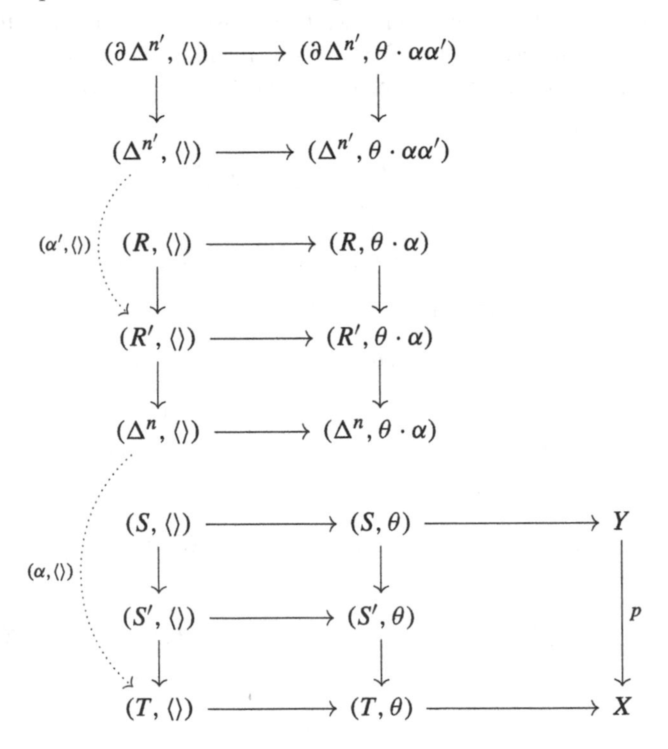

So if we have a map $(T, \langle\rangle) \to Y$ and we wish to push it forward to $(T, \theta) \to Y$, then we can do this in two steps: first we can compute $(S', \theta) \to Y$ and then compute $(T, \theta) \to Y$. The former we can compute by finding the lift $(\Delta^{n'}, \theta \cdot \alpha\alpha') \to Y$ and then pushing it down, which also can be done in two steps, yielding a map $(R', \theta \cdot \alpha) \to Y$ and then a map $(S', \theta) \to Y$. Given this map $(S', \theta) \to Y$, the induction hypothesis tells us that we can compute the desired map $(T, \theta) \to Y$ from the map $(R', \theta \cdot \alpha) \to Y$ we computed along the way by taking its induced lift $(\Delta^n, \theta \cdot \alpha) \to Y$ and then pushing that down. In other words, we take the induced map $(\Delta^n, \theta \cdot \alpha) \to Y$ and then push that down, thus showing the induction step. □

Lemma 12.6 *If all horn squares are $\mathcal{S}$-stable, then so are all one-step mould squares starting from the empty traversal.*

Proof Suppose we have a one-step mould square starting from the empty traversal, like

$$\begin{array}{ccc} (S, \langle\rangle) & \longrightarrow & (S, \langle i, \pm\rangle) \\ \downarrow & & \downarrow \\ (T, \langle\rangle) & \longrightarrow & (T, \langle i, \pm\rangle) \end{array}$$

which we want to pull it back along some map, say $(\alpha, \langle\rangle)$ with $\alpha \in \mathcal{S}$ (note that the second component has to be the empty traversal). Let $\beta : [m] \to [n]$ be the m-simplex which we need to add to S to obtain T, and consider the pullback:

$$\begin{array}{ccc} \Delta^{m'} & \xrightarrow{\alpha'} & \Delta^m \\ \downarrow{\scriptstyle \beta'} & \searrow^{\gamma} & \downarrow{\scriptstyle \beta} \\ \Delta^{n'} & \xrightarrow{\alpha} & \Delta^n \end{array}$$

(note that β is monic, so this pullback exists provided the images of α and β have some overlap: which we may assume without loss of generality, because otherwise $\alpha^* S = \alpha^* T$ and the stability condition is trivially satisfied). Note that $\alpha' \in \mathcal{S}$.

We again get four mould squares, where the dotted arrows again indicate some base changes:

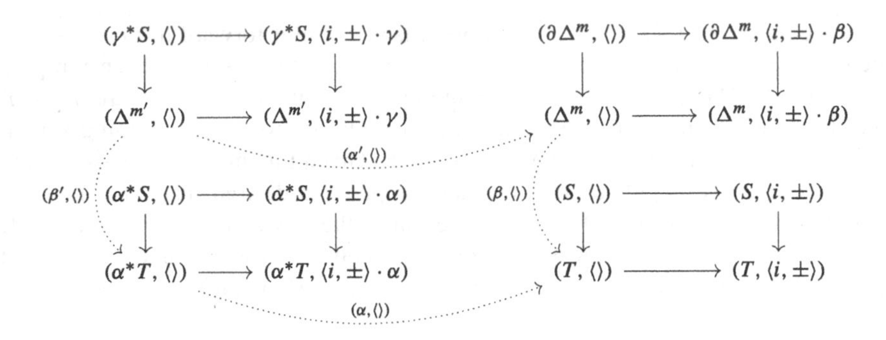

Note that the base change cubes for the arrows going down have pushouts as their left and right faces. We are given a map $(T, \langle\rangle) \to Y$ and asked to compare the induced maps $(T, \langle i, \pm\rangle) \to Y$ and $(\alpha^*T, \langle i, \pm\rangle \cdot \alpha) \to Y$. The previous lemma tells us that both induced maps can be computed by first taking the induced maps $(\Delta^{m'}, \langle i, \pm\rangle \cdot \gamma) \to Y$ and $(\Delta^m, \langle i, \pm\rangle \cdot \beta) \to Y$ and then pushing them down. So if the square on the top right is $\mathcal{S}$-stable, then so is the square on the bottom right. □

From Lemmas 12.4 and 12.6 we deduce:

Proposition 12.1 *If we equip a map $p : Y \to X$ with lifts against horn squares which are $\mathcal{S}$-stable, then all the induced lifts against small mould squares will be $\mathcal{S}$-stable.*

What does this mean for effective Kan fibrations? To equip a map $p : Y \to X$ with the structure of an effective Kan fibration, it will be (necessary and) sufficient to find lifts against horn squares so that the induced lifts against small mould squares are $\mathcal{S}$-stable for both $\mathcal{S} = \{$ face maps $\} \cup \{$ identities$\}$ and $\mathcal{S} = \{$ degeneracy maps $\} \cup \{$ identities$\}$. So the proposition tells us that we need to find lifts against horn squares which are stable relative to both classes. But lifts against horn squares are always stable relative to the first class (faces plus identities), because $d_i^* \partial \Delta^n$ is always the maximal sieve. So we have:

Theorem 12.1 *The following notions of fibred structure are isomorphic:*

- *Being an effective Kan fibration.*
- *To assign to each map all systems of lifts against horn squares which are stable along degeneracy maps.*

Remark 12.3 It should be clear that the second notion of fibred structure also naturally arises as the vertical maps in a discretely fibred concrete double category. Then this concrete double category is isomorphic to the one of given by the effective Kan fibrations: indeed, fullness on squares can be shown as in Lemma 12.1.

Let us now try to unwind what that means concretely: lifts for horn squares which are stable along degeneracies map. First of all, for each n there are $2(n + 1)$ horn squares as follows:

$$\begin{array}{ccc}
(\partial\Delta^n, \langle\rangle) & \longrightarrow & (\partial\Delta^n, \langle i, \pm\rangle) \\
\downarrow & & \downarrow \\
(\Delta^n, \langle\rangle) & \longrightarrow & (\Delta^n, \langle i, \pm\rangle),
\end{array}$$

and these can be pulled back along $s_j : \Delta^{n+1} \to \Delta^n$. The case where $j = i$ is special and we will postpone discussion of that case.

In case $j \neq i$ we have the following cartesian morphism of HDRs:

$$\begin{array}{ccccc}
\Delta^{n+1} & \xrightarrow{d_{i^*}/d_{i^*+1}} & \Delta^{n+2} & \xrightarrow{s_{i^*}} & \Delta^{n+1} \\
\downarrow{\scriptstyle s_j} & & \downarrow{\scriptstyle s_{j^*}} & & \downarrow{\scriptstyle s_j} \\
\Delta^n & \xrightarrow{d_i/d_{i+1}} & \Delta^{n+1} & \xrightarrow{s_i} & \Delta^n
\end{array}$$

where $i^* = i + 1$ if $j < i$ and $i^* = i$ if $j > i$, while $j^* = j$ if $j < i$ and $j^* = j + 1$ if $j > i$. This means that if we pull back the horn square above along s_j, then we obtain a vertical composition of three one-step mould squares, as follows:

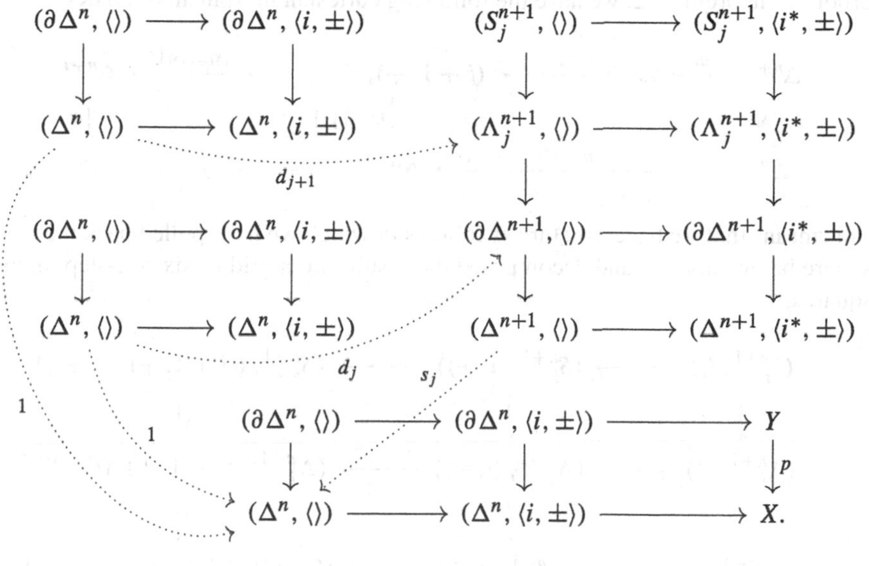

$$\begin{array}{ccccccc}
(\partial\Delta^n, \langle\rangle) & \longrightarrow & (\partial\Delta^n, \langle i, \pm\rangle) & & (S^{n+1}_j, \langle\rangle) & \longrightarrow & (S^{n+1}_j, \langle i^*, \pm\rangle) \\
\downarrow & & \downarrow & & \downarrow & & \downarrow \\
(\Delta^n, \langle\rangle) & \longrightarrow & (\Delta^n, \langle i, \pm\rangle) & \overset{d_{j+1}}{\cdots\cdots\rightarrow} & (\Lambda^{n+1}_j, \langle\rangle) & \longrightarrow & (\Lambda^{n+1}_j, \langle i^*, \pm\rangle) \\
 & & & & \downarrow & & \downarrow \\
(\partial\Delta^n, \langle\rangle) & \longrightarrow & (\partial\Delta^n, \langle i, \pm\rangle) & & (\partial\Delta^{n+1}, \langle\rangle) & \longrightarrow & (\partial\Delta^{n+1}, \langle i^*, \pm\rangle) \\
\downarrow & & \downarrow & & \downarrow & & \downarrow \\
(\Delta^n, \langle\rangle) & \longrightarrow & (\Delta^n, \langle i, \pm\rangle) & \overset{d_j}{\cdots\cdots\rightarrow} & (\Delta^{n+1}, \langle\rangle) & \longrightarrow & (\Delta^{n+1}, \langle i^*, \pm\rangle) \\
 & & & & \vdots\, {\scriptstyle s_j} & & \\
 & & (\partial\Delta^n, \langle\rangle) & \longrightarrow & (\partial\Delta^n, \langle i, \pm\rangle) & \longrightarrow & Y \\
 & & \downarrow & & \downarrow & & \downarrow{\scriptstyle p} \\
 & & (\Delta^n, \langle\rangle) & \longrightarrow & (\Delta^n, \langle i, \pm\rangle) & \longrightarrow & X.
\end{array}$$

(In the diagram, dotted arrows labelled 1 run from the two copies of $(\Delta^n, \langle\rangle)$ on the left to the bottom $(\Delta^n, \langle\rangle)$, and a dotted arrow s_j runs from $(\Delta^{n+1}, \langle\rangle)$ to the bottom $(\Delta^n, \langle\rangle)$.)

Here we have used the abbreviation $S^{n+1}_j := s^*_j \partial\Delta^n = \Lambda^{n+1}_{j,j+1} \cup (d_j \cap d_{j+1})$ (that is, it is Δ^{n+1} with the interior as well as the jth and $(j + 1)$st faces missing). Note that our recipe for computing the lifts against the first two squares in the vertical composition on the top right in the diagram tells us how to solve the original lifting problem (because $s_j.d_j = s_j.d_{j+1} = 1$). In other words, we can phrase the compatibility condition for the case $j \neq i$ as follows: if $f : (\Delta^n, < i, \pm >) \to Y$ is our chosen solution to the lifting problem

$$\begin{array}{ccccc}
(\partial\Delta^n, \langle\rangle) & \longrightarrow & (\partial\Delta^n, \langle i, \pm\rangle) & \xrightarrow{y} & Y \\
\downarrow & & \downarrow & \overset{g}{\nearrow} & \downarrow p \\
(\Delta^n, \langle\rangle) & \longrightarrow & (\Delta^n, \langle i, \pm\rangle) & \xrightarrow{x} & X,
\end{array}$$

then our chosen solution $(\Delta^{n+1}, < i^*, \pm >) \to Y$ to the lifting problem

$$\begin{array}{ccccc}
(\partial\Delta^{n+1}, \langle\rangle) & \longrightarrow & (\partial\Delta^{n+1}, \langle i^*, \pm\rangle) & \xrightarrow{y'} & Y \\
\downarrow & & \downarrow & \overset{g.s_j}{\nearrow} & \downarrow p \\
(\Delta^{n+1}, \langle\rangle) & \longrightarrow & (\Delta^{n+1}, \langle i^*, \pm\rangle) & \xrightarrow{x.s_{j^*}} & X
\end{array}$$

should be $f.s_{j^*}$, where y' is the map which is $y \cdot s_{j^*}$ on S_j^{n+1} and f on both d_j and d_{j+1}.

The case $j = i$ is special, because then the pullback of the horn square grows in a horizontal direction as well. In this case it will be convenient to treat the positive and negative case separately. So let do the positive case first. As we have seen in the proof of Theorem 10.2, we have the following cartesian morphism of HDRs:

$$\begin{array}{ccccccc}
\Delta^{n+1} & \xrightarrow{d_i} & \Delta^{n+2} & \xrightarrow{\iota_2} & < (i+1,+), (i,+) > & \xrightarrow{[s_{i+1}, s_i]} & \Delta^{n+1} \\
\downarrow s_i & & & & \downarrow [s_i, s_{i+1}] & & \downarrow s_i \\
\Delta^n & & \xrightarrow{d_i} & & \Delta^{n+1} & \xrightarrow{s_i} & \Delta^n.
\end{array}$$

This means that we have a picture as follows in which we have pulled back the horn square below along s_i and decomposed the result into a grid of six one-step mould squares.

$$\begin{array}{ccccc}
(S_i^{n+1}, \langle\rangle) & \longrightarrow & (S_i^{n+1}, \langle i, +\rangle) & \longrightarrow & (S_i^{n+1}, \langle (i+1,+), (i,+)\rangle) \\
\downarrow & & \downarrow & & \downarrow \\
(\Lambda_i^{n+1}, \langle\rangle) & \longrightarrow & (\Lambda_i^{n+1}, \langle i, +\rangle) & \longrightarrow & (\Lambda_i^{n+1}, \langle (i+1,+), (i,+)\rangle) \\
\downarrow & & \downarrow & & \downarrow \\
(\partial\Delta^{n+1}, \langle\rangle) & \longrightarrow & (\partial\Delta^{n+1}, \langle i, +\rangle) & \longrightarrow & (\partial\Delta^{n+1}, \langle (i+1,+), (i,+)\rangle) \\
\downarrow & & \downarrow & & \downarrow \\
(\Delta^{n+1}, \langle\rangle) & \longrightarrow & (\Delta^{n+1}, \langle i, +\rangle) & \longrightarrow & (\Delta^{n+1}, \langle (i+1,+), (i,+)\rangle) \\
\vdots\, s_i & & & & \\
(\partial\Delta^n, \langle\rangle) & \longrightarrow & (\partial\Delta^n, \langle i, +\rangle) & \xrightarrow{y} & Y \\
\downarrow & & \downarrow & \overset{g}{\nearrow} & \downarrow p \\
(\Delta^n, \langle\rangle) & \longrightarrow & (\Delta^n, \langle i, +\rangle) & \xrightarrow{x} & X.
\end{array}$$

Let us first consider the left column in the grid above. Note that if we pull back the first square along d_{i+1}, we get the original square back, while if we pull back the second square along d_i, we get a square which trivialises in the horizontal direction. For that reason the left column gives us the following compatibility condition: if $f : (\Delta^n, < i, + >) \to Y$ is our chosen solution to the lifting problem

$$\begin{array}{ccccc} (\partial\Delta^n, \langle\rangle) & \longrightarrow & (\partial\Delta^n, \langle i, +\rangle) & \xrightarrow{y} & Y \\ \downarrow & g & \downarrow & & \downarrow p \\ (\Delta^n, \langle\rangle) & \longrightarrow & (\Delta^n, \langle i, +\rangle) & \xrightarrow{x} & X, \end{array}$$

then our chosen solution $(\Delta^{n+1}, < i, + >) \to Y$ to the lifting problem

$$\begin{array}{ccccc} (\partial\Delta^{n+1}, \langle\rangle) & \longrightarrow & (\partial\Delta^{n+1}, \langle i, +\rangle) & \xrightarrow{y'} & Y \\ \downarrow & g.s_i & \downarrow & & \downarrow p \\ (\Delta^{n+1}, \langle\rangle) & \longrightarrow & (\Delta^{n+1}, \langle i, +\rangle) & \xrightarrow{x.s_{i+1}} & X \end{array}$$

should be $f.s_{i+1}$, where y' is the map which is $y \cdot s_{i+1}$ on S_i^{n+1}, f on d_{i+1} and g on d_i.

We now turn to the column on the right. Note that if we pull back the first square along d_{i+1}, the square trivialises in the horizontal direction, while if we pull back the second square along d_i, we get the original horn square back. Therefore the right hand column gives us the following compatibility condition: if $f : (\Delta^n, < i, + >) \to Y$ is our chosen solution to the lifting problem

$$\begin{array}{ccccc} (\partial\Delta^n, \langle\rangle) & \longrightarrow & (\partial\Delta^n, \langle i, +\rangle) & \xrightarrow{y} & Y \\ \downarrow & g & \downarrow & & \downarrow p \\ (\Delta^n, \langle\rangle) & \longrightarrow & (\Delta^n, \langle i, +\rangle) & \xrightarrow{x} & X, \end{array}$$

then our chosen solution to the lifting problem

$$\begin{array}{ccccc} (\partial\Delta^{n+1}, \langle\rangle) & \longrightarrow & (\partial\Delta^{n+1}, \langle i+1, +\rangle) & \xrightarrow{y'} & Y \\ \downarrow & f & \downarrow & & \downarrow p \\ (\Delta^{n+1}, \langle\rangle) & \longrightarrow & (\Delta^{n+1}, \langle i+1, +\rangle) & \xrightarrow{x.s_i} & X \end{array}$$

should be $f.s_i$, where y' is the map which is $y \cdot s_i$ on S_i^{n+1}, f on d_i and $f.d_{i+1}$ on d_{i+1}.

Now let us do the negative case. In that case we have following cartesian morphism of HDRs:

$$
\begin{array}{ccccccc}
\Delta^{n+1} & \xrightarrow{d_{i+2}} & \Delta^{n+2} & \xrightarrow{\iota_2} & <(i,-),(i+1,-)> & \xrightarrow{[s_i,s_{i+1}]} & \Delta^{n+1} \\
\downarrow{\scriptstyle s_i} & & & & \downarrow{\scriptstyle [s_{i+1},s_i]} & & \downarrow{\scriptstyle s_i} \\
\Delta^n & & \xrightarrow{\quad d_{i+1} \quad} & & \Delta^{n+1} & \xrightarrow{\quad s_i \quad} & \Delta^n.
\end{array}
$$

The situation we now have to look at is the one where we pull the horn square at the bottom of diagram below back along s_i and decompose the result into six one-step mould squares.

$$
\begin{array}{ccccc}
(S_i^{n+1},\langle\rangle) & \longrightarrow & (S_i^{n+1},\langle i+1,-\rangle) & \longrightarrow & (S_i^{n+1},\langle(i,-),(i+1,-)\rangle) \\
\downarrow & & \downarrow & & \downarrow \\
(\Lambda_i^{n+1},\langle\rangle) & \longrightarrow & (\Lambda_i^{n+1},\langle i+1,-\rangle) & \longrightarrow & (\Lambda_i^{n+1},\langle(i,-),(i+1,-)\rangle) \\
\downarrow & & \downarrow & & \downarrow \\
(\partial\Delta^{n+1},\langle\rangle) & \longrightarrow & (\partial\Delta^{n+1},\langle i+1,-\rangle) & \longrightarrow & (\partial\Delta^{n+1},\langle(i,-),(i+1,-)\rangle) \\
\downarrow & & \downarrow & & \downarrow \\
(\Delta^{n+1},\langle\rangle) & \longrightarrow & (\Delta^{n+1},\langle i+1,-\rangle) & \longrightarrow & (\Delta^{n+1},\langle(i,-),(i+1,-)\rangle) \\
\vdots{\scriptstyle s_i} & & & & \\
(\partial\Delta^n,\langle\rangle) & \longrightarrow & (\partial\Delta^n,\langle i,-\rangle) & \longrightarrow & Y \\
\downarrow & & \downarrow & & \downarrow{\scriptstyle p} \\
(\Delta^n,\langle\rangle) & \longrightarrow & (\Delta^n,\langle i,-\rangle) & \longrightarrow & X.
\end{array}
$$

Both columns in the grid will again determine a compatibility condition and to see what they are, we start off by considering the left hand column. Note that the pullback of the first square along d_{i+1} is trivial in the horizontal direction, while if we pull back the second square along d_i, we get our original horn square back. So the compatibility condition becomes this: if $f : (\Delta^n, < i, - >) \to Y$ is our chosen solution to the lifting problem

$$
\begin{array}{ccccc}
(\partial\Delta^n,\langle\rangle) & \longrightarrow & (\partial\Delta^n,\langle i,-\rangle) & \xrightarrow{y} & Y \\
\downarrow & {\scriptstyle g} & \downarrow & & \downarrow{\scriptstyle p} \\
(\Delta^n,\langle\rangle) & \longrightarrow & (\Delta^n,\langle i,-\rangle) & \xrightarrow{x} & X,
\end{array}
$$

then our chosen solution to the lifting problem

$$\begin{array}{ccccc} (\partial\Delta^{n+1}, \langle\rangle) & \longrightarrow & (\partial\Delta^{n+1}, \langle i+1, -\rangle) & \xrightarrow{y'} & Y \\ \downarrow & {\scriptstyle g.s_i} & \downarrow & & \downarrow p \\ (\Delta^{n+1}, \langle\rangle) & \longrightarrow & (\Delta^{n+1}, \langle i+1, -\rangle) & \xrightarrow{x.s_i} & X \end{array}$$

should be $f.s_i$, where y' is the map which is $y \cdot s_i$ on S_i^{n+1}, f on d_i and g on d_{i+1}.

Finally, if we consider the right hand column, then the pullback of the first square along d_{i+1} gives us the original horn square back, while the pullback of the second square along d_i trivialises in the horizontal direction. Therefore this column yields the following compatibility condition: if f is our chosen solution to the lifting problem

$$\begin{array}{ccccc} (\partial\Delta^{n}, \langle\rangle) & \longrightarrow & (\partial\Delta^{n}, \langle i, -\rangle) & \xrightarrow{y} & Y \\ \downarrow & {\scriptstyle g} & \downarrow & & \downarrow p \\ (\Delta^{n}, \langle\rangle) & \longrightarrow & (\Delta^{n}, \langle i, -\rangle) & \xrightarrow{x} & X, \end{array}$$

then our chosen solution to the lifting problem

$$\begin{array}{ccccc} (\partial\Delta^{n+1}, \langle\rangle) & \longrightarrow & (\partial\Delta^{n+1}, \langle i, -\rangle) & \xrightarrow{y'} & Y \\ \downarrow & {\scriptstyle f} & \downarrow & & \downarrow p \\ (\Delta^{n+1}, \langle\rangle) & \longrightarrow & (\Delta^{n+1}, \langle i, -\rangle) & \xrightarrow{x.s_{i+1}} & X \end{array}$$

should be $f.s_{i+1}$, where y' is the map which is $y \cdot s_{i+1}$ on S_i^{n+1}, f on d_{i+1} and $f.d_i$ on d_i.

To summarise the entire discussion, let us write for each $n \in \mathbb{N}$:

$$\begin{aligned} \mathcal{A}_n &= \{(i, j, i+1, j) : i, j \leq n, j < i\} \\ &\cup \{(i, j, i, j+1) : i, j \leq n, j > i\} \end{aligned}$$

Theorem 12.2 *The following notions of fibred structure are isomorphic:*

- *To be an effective Kan fibration.*
- *To assign to a map $p : Y \to X$ lifts against horn squares, in such a way that:*

(a) for any $n \in \mathbb{N}$, $(i, j, i^*, j^*) \in \mathcal{A}_n$ and $\pm \in \{+, -\}$: if f is our chosen solution to the lifting problem

$$\begin{array}{ccccc} (\partial\Delta^{n}, \langle\rangle) & \longrightarrow & (\partial\Delta^{n}, \langle i, \pm\rangle) & \xrightarrow{y} & Y \\ \downarrow & {\scriptstyle g} & \downarrow & & \downarrow p \\ (\Delta^{n}, \langle\rangle) & \longrightarrow & (\Delta^{n}, \langle i, \pm\rangle) & \xrightarrow{x} & X, \end{array}$$

then our chosen solution to the lifting problem

$$\begin{array}{ccccc} (\partial\Delta^{n+1}, \langle\rangle) & \longrightarrow & (\partial\Delta^{n+1}, \langle i^*, \pm\rangle) & \xrightarrow{y'} & Y \\ \downarrow & g.s_j & \downarrow & & \downarrow p \\ (\Delta^{n+1}, \langle\rangle) & \longrightarrow & (\Delta^{n+1}, \langle i^*, \pm\rangle) & \xrightarrow{x.s_{j^*}} & X \end{array}$$

should be $f.s_{j^*}$, where y' is the map which is $y \cdot s_{j^*}$ on S_j^{n+1} and f on the faces d_j and d_{j+1}.

(b) *for any $n \in \mathbb{N}$, if $(i, i^*, j^*) = (i, i, i+1)$ and $\pm = +$ or $(i, i^*, j^*) = (i, i+1, i)$ and $\pm = -$: if f is our chosen solution to the lifting problem*

$$\begin{array}{ccccc} (\partial\Delta^{n}, \langle\rangle) & \longrightarrow & (\partial\Delta^{n}, \langle i, \pm\rangle) & \xrightarrow{y} & Y \\ \downarrow & g & \downarrow & & \downarrow p \\ (\Delta^{n}, \langle\rangle) & \longrightarrow & (\Delta^{n}, \langle i, \pm\rangle) & \xrightarrow{x} & X, \end{array}$$

then our chosen solution to the lifting problem

$$\begin{array}{ccccc} (\partial\Delta^{n+1}, \langle\rangle) & \longrightarrow & (\partial\Delta^{n+1}, \langle i^*, \pm\rangle) & \xrightarrow{y'} & Y \\ \downarrow & g.s_i & \downarrow & & \downarrow p \\ (\Delta^{n+1}, \langle\rangle) & \longrightarrow & (\Delta^{n+1}, \langle i^*, \pm\rangle) & \xrightarrow{x.s_{j^*}} & X \end{array}$$

should be $f.s_{j^*}$, where y' is the map which is $y \cdot s_{j^*}$ on S_i^{n+1}, g on the face d_{i^*} and f on the face d_{j^*}.

(c) *for any $n \in \mathbb{N}$, if $(i, i^*, j^*) = (i, i+1, i)$ and $\pm = +$ or $(i, i^*, j^*) = (i, i, i+1)$ and $\pm = -$: if f is our chosen solution to the lifting problem*

$$\begin{array}{ccccc} (\partial\Delta^{n}, \langle\rangle) & \longrightarrow & (\partial\Delta^{n}, \langle i, \pm\rangle) & \xrightarrow{y} & Y \\ \downarrow & g & \downarrow & & \downarrow p \\ (\Delta^{n}, \langle\rangle) & \longrightarrow & (\Delta^{n}, \langle i, \pm\rangle) & \xrightarrow{x} & X, \end{array}$$

then our chosen solution to the lifting problem

$$\begin{array}{ccccc} (\partial\Delta^{n+1}, \langle\rangle) & \longrightarrow & (\partial\Delta^{n+1}, \langle i^*, \pm\rangle) & \xrightarrow{y'} & Y \\ \downarrow & f & \downarrow & & \downarrow p \\ (\Delta^{n+1}, \langle\rangle) & \longrightarrow & (\Delta^{n+1}, \langle i^*, \pm\rangle) & \xrightarrow{x.s_{j^*}} & X \end{array}$$

should be $f.s_{j^*}$, where y' is the map which is $y \cdot s_{j^*}$ on S_i^{n+1}, $f.d_{i^*}$ on the face d_{i^*} and f on the face d_{j^*}.

Remark 12.4 The second bullet in the theorem above is really a lifting condition against horn squares, not horns inclusions. We have seen at the beginning of the section that every horn inclusion is induced by some horn square, but for inner horns this horn square is not unique (it is for outer horns). Indeed, if Λ^n_i is an inner horn, then it is induced by two horn squares, one coming from $(i-1, +)$ and one coming from $(i, -)$, and our notion of an effective Kan fibration may choose different lifts for these two horn squares.

Remark 12.5 The isomorphism above can again be upgraded to one of discretely fibred concrete double categories in a manner which we will not make precise here.

A result similar to Theorem 12.2 holds for effective right (left) fibrations: this notion of fibred structure is equivalent to having the right lifting property with respect to horn squares with positive (negative) orientation, with the lifts satisfying the compatibility in the second item of that theorem. In fact, since there are no compatibility conditions relating the horn squares with different polarity, we obtain:

Proposition 12.2 *In the category of notions of fibred structure being an effective Kan fibration is the categorical product of being a effective left fibration and being an effective right fibration.*

12.2 Local Character and Classical Correctness

From the characterisation of effective Kan fibrations in Theorem 12.2 we can deduce that our notion of being an effective Kan fibration is both local and classically correct.

Corollary 12.1 *The notion of being an effective Kan fibration is a local notion of fibred structure.*

Proof Suppose $p : Y \to X$ is a map and for every pullback of p along a map $x : \Delta^n \to X$ we have a stable choice of a structure as in Theorem 12.2. Then, if we are given a lifting problem as in

$$\begin{array}{ccccc}
(\partial\Delta^n, \langle\rangle) & \longrightarrow & (\partial\Delta^n, \langle i, \pm\rangle) & \xrightarrow{y} & Y \\
\downarrow & {\scriptstyle g} & \downarrow & & \downarrow{\scriptstyle p} \\
(\Delta^n, \langle\rangle) & \longrightarrow & (\Delta^n, \langle i, \pm\rangle) & \xrightarrow{x} & X,
\end{array}$$

then we may pull p back along x and we get:

$$\begin{array}{ccccccc}
(\partial\Delta^n,\langle\rangle) & \longrightarrow & (\partial\Delta^n,\langle i,\pm\rangle) & \xrightarrow{y} & Y_x & \longrightarrow & Y \\
\downarrow & g & \downarrow & & \downarrow{\scriptstyle x^*p} & & \downarrow{\scriptstyle p} \\
(\Delta^n,\langle\rangle) & \longrightarrow & (\Delta^n,\langle i,\pm\rangle) & \xrightarrow{1} & \Delta^{n+1} & \xrightarrow{x} & X.
\end{array}$$

So, using the lifting structure of x^*p, we obtain a map $(\Delta^n, < i, \pm >) \to Y_x$ which we may compose with $Y_x \to Y$. In this way we obtain a lift against p. We still have to check that such lifts satisfy the conditions in Theorem 12.2. We only check the condition in (a), as the verification for the conditions (b) and (c) is similar.

So imagine that we wish to solve

$$\begin{array}{ccccc}
(\partial\Delta^{n+1},\langle\rangle) & \longrightarrow & (\partial\Delta^{n+1},\langle i^*,\pm\rangle) & \xrightarrow{y'} & Y \\
\downarrow & g.s_j & \downarrow & & \downarrow{\scriptstyle p} \\
(\Delta^{n+1},\langle\rangle) & \longrightarrow & (\Delta^{n+1},\langle i^*,\pm\rangle) & \xrightarrow{x.s_{j^*}} & X
\end{array}$$

with $(i, j, i^*, j^*) \in \mathcal{A}_n$ and where y' is the map which is $y \cdot s_{j^*}$ on S_j^{n+1} and f on the faces d_j and d_{j+1}. The recipe we were given is that we write this as

$$\begin{array}{ccccccc}
(\partial\Delta^{n+1},\langle\rangle) & \longrightarrow & (\partial\Delta^{n+1},\langle i^*,\pm\rangle) & \xrightarrow{y'} & Y_{x.s_{j^*}} & \longrightarrow & Y \\
\downarrow & g.s_j & \downarrow & & \downarrow{\scriptstyle (x.s_{j^*})^*p} & & \downarrow{\scriptstyle p} \\
(\Delta^{n+1},\langle\rangle) & \longrightarrow & (\Delta^{n+1},\langle i^*,\pm\rangle) & \xrightarrow{1} & \Delta^{n+2} & \xrightarrow{x.s_{j^*}} & X,
\end{array}$$

find the induced lift $(\Delta^{n+1}, < i^*, + >) \to Y_{x.s_{j^*}}$ and compose with $Y_{x.s_{j^*}} \to Y$. But we may write the pullback in the previous diagram as the composition of two pullbacks, as follows:

$$\begin{array}{ccccccccc}
(\partial\Delta^{n+1},\langle\rangle) & \longrightarrow & (\partial\Delta^{n+1},\langle i^*,\pm\rangle) & \xrightarrow{y'} & Y_{x.s_{j^*}} & \longrightarrow & Y_x & \longrightarrow & Y \\
\downarrow & g.s_j & \downarrow & & \downarrow{\scriptstyle (x.s_{j^*})^*p} & & \downarrow{\scriptstyle x^*p} & & \downarrow{\scriptstyle p} \\
(\Delta^{n+1},\langle\rangle) & \longrightarrow & (\Delta^{n+1},\langle i^*,\pm\rangle) & \xrightarrow{1} & \Delta^{n+2} & \xrightarrow{s_{j^*}} & \Delta^{n+1} & \xrightarrow{x} & X,
\end{array}$$

By our stability assumption, this means that the composition of the induced lift $(\Delta^{n+1}, < i^*, + >) \to Y_{x.s_{j^*}}$ with $Y_{x.s_{j^*}} \to Y_x$ is the induced lift $(\Delta^{n+1}, < i^*, + >) \to Y_x$ against x^*p. But the latter is $f.s_{j^*}$, because x^*p has lifts satisfying the condition in Theorem 12.2. This means that p has lifts satisfying that condition as well, finishing the proof. □

Corollary 12.2 *In a classical metatheory, every map which has the right lifting property against horns (a Kan fibration in the usual sense) can be equipped with the structure of an effective Kan fibration.*

Proof Suppose that we have a map which has the right lifting property against all horns. Because a lifting problem against a horn has at most one degenerate solution (see Proposition C.1), we may always choose the degenerate solution if it exists. In that case our lifts will satisfy the condition in Theorem 12.2, because it says that under certain circumstances we should choose a degenerate solution. But by always choosing the unique degenerate solution (if it exists), this will automatically be satisfied. □

Remark 12.6 We again have similar results for effective left and right fibrations. Indeed, proofs which are almost identical to the ones of Corollaries 12.1 and 12.2 yield:

- Being an effective right (left) fibration is a local notion of fibred structure.
- In a classical metatheory, a map can be equipped with the structure of an effective right (left) fibration if and only if it has the right lifting property against horn inclusions $\Lambda^n_i \to \Delta^n$ with $i \neq 0$ (with $i \neq n$), that is, if and only if it is a right (left) fibration in the usual sense.

In addition, the definition of effective Kan fibrations in terms of horn squares allows us to prove that they have the following properties:

Corollary 12.3 *Effective Kan fibrations have the following properties:*

1. *If each of $f_i : A_i \to B_i$ is effectively Kan, then so is their sum $\coprod f_i : \coprod A_i \to \coprod_{i \in I} B_i$.*
2. *If each of $f_i : A_i \to B$ is effectively Kan, then so is their copairing $\coprod f_i : \coprod A_i \to B$.*
3. *Discrete presheaves, such as the terminal object and the natural numbers object $\mathbb{N}$, are effectively Kan.*

Proof Points (1) and (2) follow from Theorem 12.2 in combination with the fact that any map $\Delta^n \to \coprod_{i \in I} A_i$ factors through one of the coproduct inclusions $A_j \to \coprod_{i \in I} A_i$.

Point (3) follows immediately from point (2) since discrete presheaves are coproducts $\coprod_{x \in X} 1$ of the terminal object. □

Chapter 13
Conclusion

In this final chapter we would like to take stock of the properties of effective Kan fibrations that we have established and outline some directions for future research.

13.1 Properties of Effective Kan Fibrations

The properties of effective Kan fibrations that we have established in this book are:

1. Effective Kan fibrations are closed under composition and pullback.
2. Isomorphisms are effective Kan fibrations.
3. If $f : Z \to Y$ and $g : Y \to X$ are effective Kan fibrations, then so is $g_* f$, the pushforward of f along g.
4. The notion of an effective Kan fibration is local and hence universal effective Kan fibrations exist.
5. The notion of an effective Kan fibration is classically correct in that, in a classical metatheory, a map of simplicial sets can be equipped with the structure of an effective Kan fibration precisely when it has the right lifting property against horn inclusions.
6. If each of $f_i : A_i \to B$ is effectively Kan, then so is their copairing $\coprod f_i : \coprod A_i \to B$. In particular, discrete presheaves, such as the terminal object and the natural numbers object $\mathbb{N}$, are effectively Kan.

Properties (1) and (2) are immediate from the fact that effective Kan fibration are defined via a lifting property against a triple category. Property (3) was Theorem 7.1. Property (4) was Corollary 12.1 in combination with Theorem 2.1. Property (5) was Corollary 12.2. Property (6) was Corollary 12.3.

The main disadvantage of the notion of an effective Kan fibration is that it is not clear (constructively!) that they are closed under retracts. That is, it is unclear how

B. van den Berg, E. Faber, *Effective Kan Fibrations in Simplicial Sets*,
Lecture Notes in Mathematics 2321,
https://doi.org/10.1007/978-3-031-18900-5_13

from a diagram

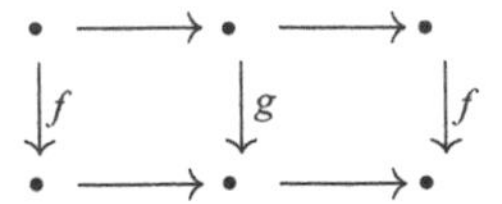

showing that f is a retract of g and an effective Kan fibration structure on g, one is supposed to construct an effective Kan fibration structure on f. (Of course, one can in a classical metatheory, because the notion of an effective Kan fibration is classically correct and the usual Kan fibrations are closed under retracts.) This leads to:

Open question: Can it be proved in a constructive metatheory that the effective Kan fibrations are closed under retracts?

13.2 Directions for Future Research

In future work we plan to show that the effective Kan fibrations are the right class in an algebraic weak factorisation system. At the point of writing we have a classical proof of this fact; we are working on transforming this into a constructive one.

This would also be the first step towards a constructive proof showing that the effective Kan fibrations and the effective trivial Kan fibration yield an algebraic model structure (as in [22]) on the category of simplicial sets.

Ultimately we hope to build a model of homotopy type theory in the category of simplicial sets based on our notion of an effective Kan fibration. Here the main challenge will be to construct univalent universes. Other structure that one might hope to be present are certain inductive types, like W-types, as in [46]. All of this is left to future work.

Appendix A
Axioms

In this appendix we will collect the axioms for a Moore category and a dominance that play a role in this book. The reader can think of these as our version of the Orton-Pitts axioms [20] (see also [8, 21]).

A.1 Moore Structure

Our first ingredient is a suitable notion of Moore paths.

Definition A.1 Let $\mathcal{E}$ be a category with finite limits. A *Moore structure* on $\mathcal{E}$ consists of the following data:

1. We have a pullback-preserving endofunctor M on $\mathcal{E}$ together with natural transformations $r : 1_{\mathcal{E}} \to M$, $s, t : M \to 1_{\mathcal{E}}$, and $\mu : M_t \times_s M \to M$ turning every object X in $\mathcal{E}$ in the object of objects of an internal category, with MX as the object of arrows. Note the order in which μ takes its arguments: it is not in the way categorical composition is usually written. The reason is that we think of μ as concatenation of paths rather than as categorical composition and we write it as such.
2. There is a natural transformation $\Gamma : M \to MM$ making (M, s, Γ) into a comonad.
3. There is a strength $\alpha_{X,Y} : X \times MY \to M(X \times Y)$, that is, α is a natural transformation making

B. van den Berg, E. Faber, *Effective Kan Fibrations in Simplicial Sets*,
Lecture Notes in Mathematics 2321, https://doi.org/10.1007/978-3-031-18900-5

$$\begin{array}{ccc}
(X \times Y) \times MZ & \xrightarrow{\alpha_{X\times Y,Z}} & M((X \times Y) \times Z) \\
\downarrow \cong & & \downarrow \cong \\
X \times (Y \times MZ) \xrightarrow[1_X \times \alpha_{Y,Z}]{} & X \times M(Y \times Z) & \xrightarrow[\alpha_{X,Y\times Z}]{} M(X \times (Y \times Z))
\end{array}$$

$$\begin{array}{ccc}
X \times MY & \xrightarrow{\alpha_{X,Y}} & M(X \times Y) \\
& {\scriptstyle p_2} \searrow & \downarrow {\scriptstyle Mp_2} \\
& & MY
\end{array}$$

commute. In addition, all the previous structure is strong, so the following diagrams commute as well:

$$\begin{array}{ccc}
X \times MY & \xrightarrow{\alpha} & M(X \times Y) \\
{\scriptstyle 1\times s}\downarrow \; \uparrow{\scriptstyle 1\times r} \; \downarrow{\scriptstyle 1\times t} & & {\scriptstyle s}\downarrow \; \uparrow{\scriptstyle r} \; \downarrow{\scriptstyle t} \\
X \times Y & \xrightarrow{1} & X \times Y
\end{array}
\qquad
\begin{array}{ccc}
X \times MY_t \times_s MY & \xrightarrow{(\alpha.(p_1,p_2),\alpha.(p_1,p_3))} & M(X \times Y)_t \times_s M(X \times Y) \\
\downarrow {\scriptstyle 1\times\mu} & & \downarrow {\scriptstyle \mu} \\
X \times MY & \xrightarrow{\alpha} & M(X \times Y)
\end{array}$$

$$\begin{array}{ccc}
X \times MY & \xrightarrow{\alpha} & M(X \times Y) \\
\downarrow {\scriptstyle 1\times\Gamma} & & \downarrow {\scriptstyle \Gamma} \\
X \times MMY \xrightarrow[\alpha]{} & M(X \times MY) & \xrightarrow[M\alpha]{} MM(X \times Y).
\end{array}$$

4. We have the following axioms for the connection Γ (interaction with r, t):

$$\Gamma.r = rM.r, \qquad tM.\Gamma = r.t, \qquad Mt.\Gamma = \theta_X.\alpha_{X,1}.(t, M!),$$

with θ_X being the iso $Mp_1 : M(X \times 1) \to MX$.

5. And, finally, we have the following *distributive law* (interaction between Γ and μ):

$$\begin{aligned}\Gamma.\mu &= \mu.(M\mu.\nu.(\Gamma.p_1, \theta_{MX}.\alpha_{MX,1}.(p_2, M!.p_1)), \Gamma.p_2) \\ &: MX \times_X MX \to MMX\end{aligned}$$

with ν being the natural transformation (in this case $MMX \times_{MX} MMX \to M(MX \times_X MX)$) induced by preservation of pullbacks. This condition can be visualized as follows. When $p, q \in MX$ are composable Moore paths as in the left-hand size of the diagram, then $\Gamma.\mu(p, q)$ is defined by $\Gamma(p)$ and $\Gamma(q)$ in the following way:

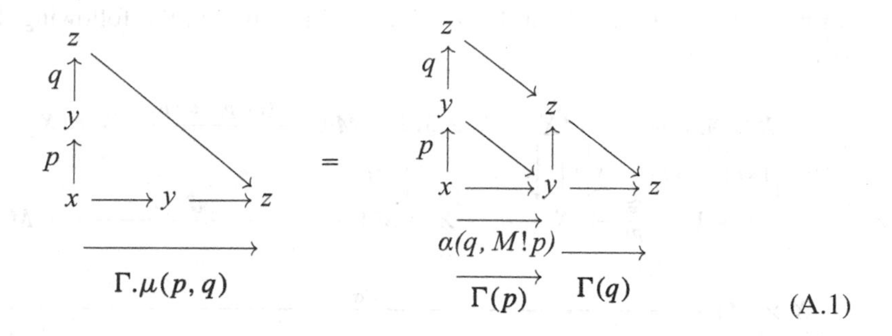

(A.1)

Whenver a category $\mathcal{E}$ is equipped with structure thus described, we call $\mathcal{E}$ a *category with Moore structure*, or a *Moore category* for short.

Remark A.1 The notion of a path object category from [19] can be obtained from this by dropping the coassociativity axiom for Γ as well as the distributive law, whilst adding a symmetry τ (see below).

Remark A.2 As observed in [19], the fact that M preserves pullbacks means that the entire strength is determined by the maps

$$\alpha_X := X \times M1 \xrightarrow{\alpha_{X,1}} M(X \times 1) \xrightarrow{\theta_X} MX.$$

The reason for this is that the outer rectangle and right hand square in

$$\begin{array}{ccccc} X \times MY & \xrightarrow{\alpha_{X,Y}} & M(X \times Y) & \xrightarrow{Mp_2} & MY \\ \downarrow{\scriptstyle 1_X \times M!} & & \downarrow{\scriptstyle Mp_1} & & \downarrow{\scriptstyle M!} \\ X \times M1 & \xrightarrow{\alpha_X} & MX & \xrightarrow{M!} & M1 \end{array}$$

(with the outer arrows $p_2 : X \times MY \to MY$ and $p_2 : X \times M1 \to M1$)

are pullbacks. And, if we wish, axioms (4-6) can also be formulated as follows: there is a natural transformation

$$\alpha_X : X \times M1 \to MX$$

with $M!.\alpha_X = p_2 : X \times M1 \to M1$, and, in addition, the following diagrams commute:

$$\begin{array}{ccc} X \times M1 & \xrightarrow{\alpha} & MX \\ {\scriptstyle 1\times s}\downarrow \; {\scriptstyle 1\times r}\uparrow \; \downarrow{\scriptstyle 1\times t} & & {\scriptstyle s}\downarrow \; {\scriptstyle r}\uparrow \; \downarrow{\scriptstyle t} \\ X \times 1 & \xrightarrow[p_1]{\cong} & X \end{array} \qquad \begin{array}{ccc} X \times M1 \times M1 & \xrightarrow{(\alpha.(p_1,p_2),\alpha.(p_1,p_3))} & MX \times_X MX \\ \downarrow{\scriptstyle 1\times\mu} & & \downarrow{\scriptstyle \mu} \\ X \times M1 & \xrightarrow{\alpha} & MX \end{array}$$

$$\begin{array}{ccccccc} X \times M1 & & & \xrightarrow{\alpha} & & & MX \\ \downarrow{\scriptstyle 1\times\Gamma} & & & & & & \downarrow{\scriptstyle \Gamma} \\ X \times MM1 & \xrightarrow[(\alpha.(p_1,M!.p_2),p_2)]{} & MX \times_{M1} MM1 & \xrightarrow{\cong} & M(X \times M1) & \xrightarrow[M\alpha]{} & MMX. \end{array}$$

Finally, we have the following axioms for the interaction between the connection Γ and the category structure:

$$\begin{aligned} \Gamma.r &= rM.r, \\ tMa.\Gamma &= r.t, \\ Mt.\Gamma &= \alpha.(t, M!), \\ \Gamma.\mu &= \mu.(M\mu.\nu_X.(\Gamma.p_1, \alpha_{MX}.(p_2, M!.p_1)), \Gamma.p_2). \end{aligned}$$

Definition A.2 We will call a Moore structure *two-sided* if it also comes equipped with a map $\Gamma^* : M \to MM$ turning (M, t, Γ^*) into a strong comonad, and such that the following equations hold:

$$\begin{aligned} \Gamma^*.r &= rM.r, \\ s.\Gamma^* &= r.s, \\ Ms.\Gamma^* &= \alpha.(s, M!), \\ \Gamma^*.\mu &= \mu.(\Gamma^*.p_1, M\mu.\nu.(\alpha_M.(p_1, M!.p_2), \Gamma^*.p_2)). \end{aligned} \tag{A.2}$$

This has the effect that if we switch s and t and define $\mu^* := \mu.(p_2, p_1)$, then we get a second Moore structure. We will also require that μ is both left and right cancellative, and that we have the sandwich equation:

$$M\mu.\nu.(\Gamma^*, \Gamma) = \alpha.(1, M!) : M \to MM$$

(which also implies $M\mu^*.\nu.(\Gamma, \Gamma^*) = \alpha.(1, M!)$).

Definition A.3 A two-sided Moore structure will be called *symmetric* if it also comes equipped with a natural transformation $\tau : M \to M$ (sometimes referred to as a 'twist map') such that

$$
\begin{aligned}
\tau.\tau &= 1,\\
\tau.r &= r,\\
s.\tau &= t,\\
t.\tau &= s,\\
\Gamma^* &= \tau M.M\tau.\Gamma.\tau,
\end{aligned}
$$

while also the following diagrams commute:

$$
\begin{array}{ccc}
X \times M1 & \xrightarrow{\alpha} & MX \\
{\scriptstyle 1\times\tau}\downarrow & & \downarrow{\scriptstyle \tau} \\
X \times M1 & \xrightarrow{\alpha} & MX
\end{array}
\qquad
\begin{array}{ccc}
MX_t \times_s MX & \xrightarrow{\mu} & MX \\
{\scriptstyle (\tau.p_2,\tau.p_1)}\downarrow & & \downarrow{\scriptstyle \tau} \\
MX_t \times_s MX & \xrightarrow{\mu} & MX
\end{array}
$$

Remark A.3 There is a bit of redundancy in the previous definition, in that $\Gamma^* = \tau M.M\tau.\Gamma.\tau$ implies the equations (A.2) above.

Example: Symmetric Moore Categories

The following examples from [19] all satisfy the axioms for a symmetric Moore category.

1. The category of topological spaces with

$$MX = \sum_{r \in \mathbb{R}_{\geq 0}} X^{[0,r]},$$

the space of Moore paths.

2. The category of small groupoids with

$$MX = X^{\mathbb{I}},$$

where $\mathbb{I}$ is the interval groupoid containing two objects and one arrow $x \to y$ for any pair of objects (x, y). In fact, this also defines a symmetric Moore structure on the category of small categories.

3. The category of chain complexes over a ring R.

For more details we refer to [19, Section 5].

A.2 Dominance

The second ingredient is a dominance[32].

Definition A.4 A *dominance* on a category $\mathcal{E}$ is a class of monomorphism Σ in $\mathcal{E}$ satisfying the following three properties:

1. every isomorphism is in Σ and Σ is closed under composition.
2. every pullback of a map in Σ again belongs to Σ.
3. the category Σ_{cart} of maps in Σ and pullback squares between them has a terminal object $1 \to \Sigma$.

For some of our arguments it will be convenient to assume the following two additional axioms:

1. The elements in Σ are closed under finite unions; that is, $0 \to X$ always belongs to Σ and whenever $A \to X$ and $B \to X$ belong to Σ, then so does $A \cup B \to X$.
2. The morphism $r_X : X \to MX$ belongs to Σ for any object X.

Appendix B
Cubical Sets

In this appendix, we show that the category of *cubical sets* ([9], also [8]) can be equipped with parts of the definition of a Moore structure. This makes it possible to highlight the relationship between the notion of Moore structure introduced in this book and the related notion of *connections* in cubical sets. The conclusion is that, for the standard choice of path object, all structure except from path composition is present. The path contraction Γ corresponds to the notion of *connection* in cubical sets.

The *category of cubes* C has as objects finite subsets I of a fixed and countable set of *names* $\{x, y, z, \dots\}$, and morphisms $I \to J$ are set maps $J \to \mathsf{dM}(I)$ into the *free de Morgan algebra* generated by I. The free de Morgan algebra is the bounded distributive lattice generated by the set of constants and binary operations $\{0, 1, \wedge, \vee\}$, and the involutive function $\neg$ representing negation, which needs to satisfy de Morgan's laws.

The category of cubical sets is the category of presheaves on C. A cubical set X can be regarded as a set $X(x_1, \dots, x_n)$ for every given tuple of variables $x_1, \dots, x_n$, whose elements support several substitution operations, e.g.:

$$x_i = 0\colon X(x_1, \dots, x_n) \to X(x_1, \dots, \widehat{x_i}, \dots, x_n) \tag{B.1}$$

The operations $\wedge, \vee$ give rise to *connections*, while the operation $\neg$ gives rise to a *symmetry*. Therefore this category is sometimes referred to as cubical sets with connections and symmetries, to distinguish it from possible other (more restricted) definitions. For an introduction to cubical sets, we refer to [9].

An element $\gamma \in X(y_1, \dots, y_n, x)$ can be regarded as a 'path' between $X(x = 0)(\gamma)$ and $X(x = 1)(\gamma)$. This suggests to define a 'path object functor' M as follows:

$$MX(I) := X(I \cup \{x\}) \text{ where } x \notin I \tag{B.2}$$

$$M\eta(I) := \eta_{I \cup \{x\}} \tag{B.3}$$

B. van den Berg, E. Faber, *Effective Kan Fibrations in Simplicial Sets*,
Lecture Notes in Mathematics 2321, https://doi.org/10.1007/978-3-031-18900-5

with source, target, and identity maps induced by:

$$s_X(I) := X(x=0)\colon X(I \cup \{x\}) \to X(I)$$
$$t_X(I) := X(x=1)\colon X(I \cup \{x\}) \to X(I)$$
$$r_X(I) := X(\mathsf{id})\colon X(I) \to X(I \cup \{x\})$$

Using connections, we can define a contraction Γ:

$$\Gamma_X(I) := X(x = x \vee y)\colon X(I \cup \{x\}) \to X(I \cup \{x\} \cup \{y\}).$$

We can readily verify that M is a comonad:

$$\begin{aligned}
s_{MX}.\Gamma_X &= X(y=0).X(x = x \vee y) = 1_{MX} \\
M(s_X).\Gamma_X &= M(X(x=0)).X(x = x \vee y) = X(y=0).X(x = x \vee y) \\
&= 1_{MX} \\
\Gamma_{MX}.\Gamma_X &= X(y = y \vee z).X(x = x \vee y) \\
&= M(\Gamma_X).\Gamma_X
\end{aligned}$$

Similarly, it is easy to define a strength and show that requirement (4) of Definition A.1 holds.

Further, the $\wedge$ operation would define a dual Γ^* as in the definition of two-sided Moore structure (Definition A.2), whereas negation $\neg$ can be used to define a 'twist map' as in Definition A.3. The fact that the Γ^* derived from the twist map is the same as the former follows from the fact that negation is required to be an involution, i.e., $\neg\neg x = x$.

All that is missing is a definition of multiplication, or path concatenation:

$$\mu_X(I) : X(I \times \{x\})_{x=1} \times_{x=0} X(I \times \{x\}) \to X(I \times \{x\}) \tag{B.4}$$

To define this, it would be needed to 'compose' a pair of paths γ, γ' which satisfy $X(x=1)(\gamma) = X(x=0)(\gamma')$. Yet, given that all paths have only single length, this is not straightforward.

Appendix C
Degenerate Horn Fillers Are Unique

The purpose of this appendix is to show that horn filling problems have at most one degenerate filler, in the following sense:

Proposition C.1 *If both $x_0 \cdot \sigma_0$ and $x_1 \cdot \sigma_1$ are fillers for*

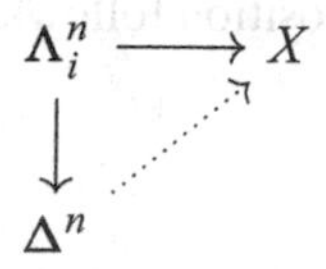

where $\sigma_0 : \Delta^n \to \Delta^k$ and $\sigma_1 : \Delta^n \to \Delta^l$ are epimorphisms in $\mathbf{\Delta}$ different from the identity, then $x_0 \cdot \sigma_0 = x_1 \cdot \sigma_1$.

The proof strategy that we will follow here was suggested to us by Christian Sattler. The (constructive) argument relies on the following lemma (see [46, Lemma 5.6]):

Lemma C.1 *Suppose that we have a diagram of the form*

$$\begin{array}{ccccc} \bullet & \longrightarrow & \bullet & \longrightarrow & \bullet \\ \downarrow{\scriptstyle f} & & \downarrow{\scriptstyle g} & & \downarrow{\scriptstyle f} \\ \bullet & \longrightarrow & \bullet & \longrightarrow & \bullet \end{array}$$

exhibiting f as a retract of g, while g has a section. Then the right hand square is an absolute pushout.

Proof *(Of Proposition C.1)* It suffices to consider the case where $\sigma_0 = s_i$ and $\sigma_1 = s_j$ with $i < j$.

Because $i \neq j + 1$ in at least one of the following diagrams the dotted arrow exists:

B. van den Berg, E. Faber, *Effective Kan Fibrations in Simplicial Sets*,
Lecture Notes in Mathematics 2321, https://doi.org/10.1007/978-3-031-18900-5

$$\begin{array}{ccccc} & & \Lambda^n_k & & \\ & & \downarrow & & \\ \Delta^{n-1} & \xrightarrow{d_i} & \Delta^n & \xrightarrow{s_i} & \Delta^{n-1} \\ \downarrow{\scriptstyle s_{j-1}} & & \downarrow{\scriptstyle s_j} & & \downarrow{\scriptstyle s_{j-1}} \\ \Delta^{n-2} & \xrightarrow{d_i} & \Delta^{n-1} & \xrightarrow{s_i} & \Delta^{n-2}. \end{array} \qquad \begin{array}{ccccc} & & \Lambda^n_k & & \\ & & \downarrow & & \\ \Delta^{n-1} & \xrightarrow{d_{j+1}} & \Delta^n & \xrightarrow{s_j} & \Delta^{n-1} \\ \downarrow{\scriptstyle s_i} & & \downarrow{\scriptstyle s_i} & & \downarrow{\scriptstyle s_i} \\ \Delta^{n-2} & \xrightarrow{d_j} & \Delta^{n-1} & \xrightarrow{s_{j-1}} & \Delta^{n-2} \end{array}$$

In either case, the previous lemma implies that both the inner and outer square in

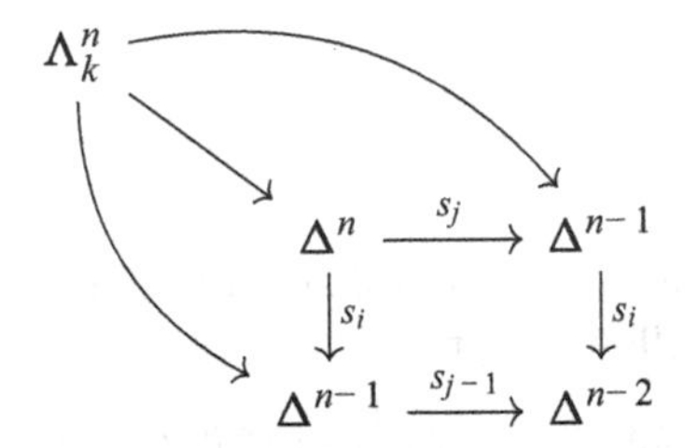

are pushouts, from which the proposition follows. □

Appendix D
Uniform Kan Fibrations

This appendix is devoted to a proof by Christian Sattler [47] showing that the notion of a uniform Kan fibration as used in [8] is not a local notion of fibred structure. In this connection it is interesting to note that their definition is the same as the one used by Coquand and others [9] for the case of cubical sets, where this notion of fibred structure is local.

Their definition makes use of the *pushout product* (or *Leibniz product*) $f \hat{\otimes} g$ of two maps $f : A \to B$ and $g : C \to D$, which is the inscribed map from the pushout as in

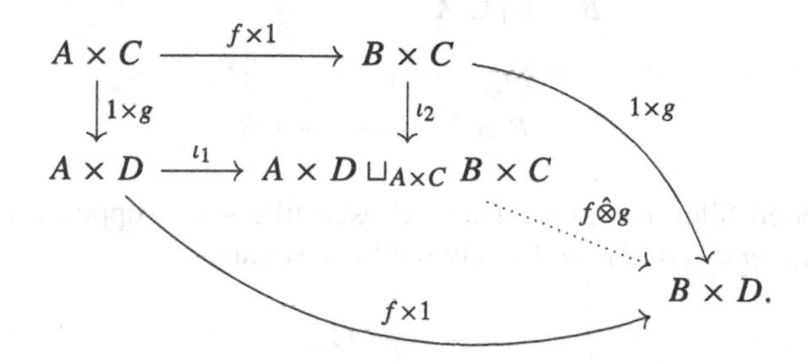

For the cofibrations they take the same class as maps as us (or as in the standard theory): the (pointwise decidable) monomorphisms. The generating trivial cofibrations are then the maps $m \hat{\otimes} \delta_i$ where m is cofibration and $\delta_i : 1 \to \mathbb{I}$ is one of the end point inclusions. It is a classical result [18, Chap. IV, Sec. 2] that these maps, like the horn inclusions, generate the Kan fibrations. In fact, this class would already be generated by the maps $m \hat{\otimes} d_i$ where d_i is an end point inclusion and m is a cofibrant sieve $S \subseteq \Delta^n$ (see *loc.cit.*).

The innovation of Gambino and Sattler is to add a uniformity condition to this definition, where this uniformity condition was suggested by the work of Coquand

B. van den Berg, E. Faber, *Effective Kan Fibrations in Simplicial Sets*,
Lecture Notes in Mathematics 2321, https://doi.org/10.1007/978-3-031-18900-5

and others. The cofibrations can be seen as the objects in a category where the morphisms $m \to n$ are pullback squares

$$\begin{array}{ccc} A & \longrightarrow & C \\ \big\downarrow{\scriptstyle m} & & \big\downarrow{\scriptstyle n} \\ B & \longrightarrow & D \end{array}$$

with composition and identities as in the arrow category on simplicial sets. If we have such a pullback square and $\delta_i : 1 \to \mathbb{I}$ is an endpoint inclusion, then we obtain another pullback square

$$\begin{array}{ccc} B \times \{i\} \cup A \times \mathbb{I} & \longrightarrow & D \times \{i\} \cup C \times \mathbb{I} \\ \big\downarrow{\scriptstyle m\hat{\otimes}d_i} & & \big\downarrow{\scriptstyle n\hat{\otimes}d_i} \\ B \times \mathbb{I} & \longrightarrow & D \times \mathbb{I}. \end{array}$$

This suggest the following notion of fibred structure on simplicial sets:

Definition D.1 [8] A *uniform Kan fibration structure* on a map $p : Y \to X$ of simplicial sets is a function which, given a cofibration $m : A \to B$, an endpoint inclusion $\delta_i : 1 \to \mathbb{I}$ and a solid commutative square

$$\begin{array}{ccc} B \times \{i\} \cup A \times \mathbb{I} & \longrightarrow & Y \\ {\scriptstyle m\hat{\otimes}\delta_i}\big\downarrow & \nearrow & \big\downarrow{\scriptstyle p} \\ B \times \mathbb{I} & \longrightarrow & X \end{array}$$

chooses a dotted filler as shown. These chosen fillers are supposed to satisfy the following *uniformity condition*: for any pullback square

$$\begin{array}{ccc} A & \longrightarrow & C \\ \big\downarrow{\scriptstyle m} & & \big\downarrow{\scriptstyle n} \\ B & \longrightarrow & D, \end{array}$$

the chosen fillers should make the inscribed triangle in

$$\begin{array}{ccccc} B \times \{i\} \cup A \times \mathbb{I} & \longrightarrow & D \times \{i\} \cup C \times \mathbb{I} & \longrightarrow & Y \\ \big\downarrow{\scriptstyle m\hat{\otimes}d_i} & & \big\downarrow{\scriptstyle n\hat{\otimes}d_i} & & \big\downarrow{\scriptstyle p} \\ B \times \mathbb{I} & \longrightarrow & D \times \mathbb{I} & \longrightarrow & X \end{array}$$

commute.

In [8] Gambino and Sattler show that their notion of uniform Kan fibration satisfies the following properties:

(1) One can show, constructively, that the uniform Kan fibrations are closed under pushforward: so if f and g are uniform Kan fibration, then so is $\Pi_f(g)$.
(2) Their notion is classically correct, in that one can show in a classical metatheory that every map which has the right lifting property against horn inclusions can be equipped with the structure of a uniform Kan fibration.

In this way they circumvent the BCP-obstruction as discussed in the introduction. What is left open, however, is whether it is also local. It turns out that it is not and in this appendix we will give Christian Sattler's proof of this fact. (It should be noted that the proof makes use of classical logic).

Remark D.1 In terms of mould squares the uniform Kan fibrations can be understood as follows. Note that $p : Y \to X$ having the right lifting property against a map $m \hat{\otimes} \delta_i$ means that given any solid diagram as in

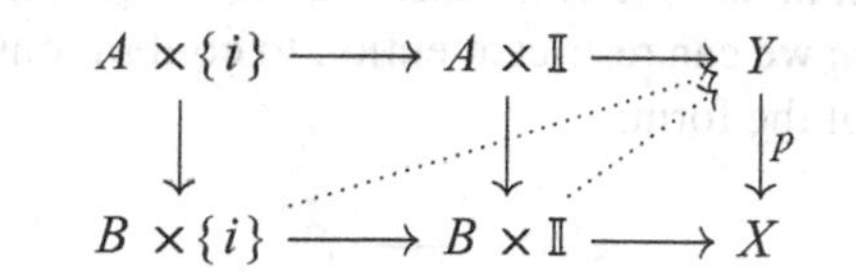

and dotted arrow $B \times \{i\} \to Y$ we can find a dotted arrow (pushforward) $B \times \mathbb{I} \to Y$ as shown. Since $\delta_i : 1 \to \mathbb{I}$ are HDRs (they are maps of the form $\Delta^n \to \widehat{\theta}$ with θ being the 1-dimensional traversal $< 0, \pm >$), the square on the left in the diagram above is a mould square. The uniformity condition says the following: suppose $m \to n$ is a pullback square and we are given a solid diagram

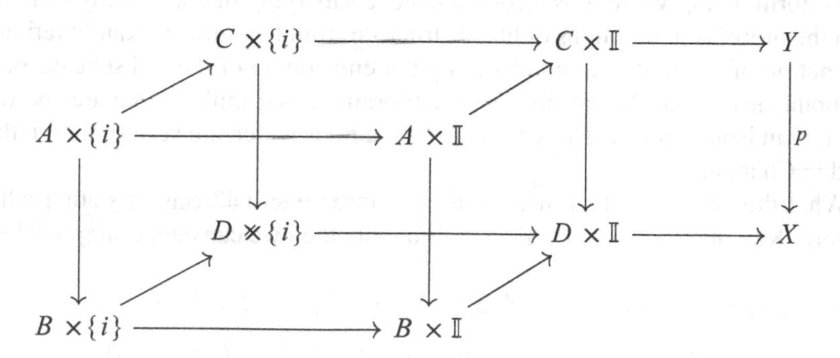

Then any arrow $D \times \{i\} \to Y$ making everything commute induces two pushforwards: one of the form $D \times \mathbb{I} \to Y$ using that the back of the cube is a mould square and one of the form $B \times \mathbb{I} \to Y$ using that the front of the cube is a mould square. We demand that these commute via the map $B \times \mathbb{I} \to D \times \mathbb{I}$. Since the cube in the diagram above is a mould cube (a cube in the triple category of mould squares), we see that this uniformity condition is also part of our definition of an effective

Kan fibration. In particular, effective Kan fibrations are uniform Kan fibrations; or, more precisely, there is a morphism of notions of fibred structure from the effective Kan fibrations to the uniform Kan fibrations. So what the notion of an effective Kan fibration adds to that of a uniform Kan fibration is the following:

(1) Instead of requiring the existence of pushforwards for HDRs of the form $B \times \{i\} \to B \times \mathbb{I}$ only, we demand their existence for all HDRs.
(2) Since HDRs can be composed, mould squares can be composed horizontally leading to an additional horizontal compatibility condition.
(3) Since cofibrations can be composed, mould squares can be composed vertically leading to an additional vertical compatibility condition.
(4) There is more general notion of mould cube leading to a more general perpendicular compatibility condition.

As we have seen, the result is a local notion of fibred structure.

In order to show that the uniform Kan fibrations are not a local notion of fibred structure, we make a number of useful observations. First of all, in the definition of a uniform Kan fibration we can restrict attention to cofibrations of the form $S \subseteq \Delta^n$ and pullback squares of the form:

$$\begin{array}{ccc} \alpha^*S & \longrightarrow & S \\ \downarrow & & \downarrow \\ \Delta^m & \xrightarrow{\alpha} & \Delta^n. \end{array}$$

To be a bit more precise: if we would define another notion of fibred structure by assiging to each map $p : Y \to X$ all the solutions to lifting problems against maps of the form $m \hat{\otimes} \delta_i$ where m is a cofibrant sieve satisfying the uniformity condition, then the morphism of notions of fibred structure from the uniform Kan fibrations to this notion of fibred structure induced by the embedding of the full subcategory of cofibrant sieves into the category of al cofibrations and pullback squares between them, is an isomorphism. This can be using methods which are very similar to those used in Chap. 8.

What this implies is that there is also a more "internal" way of stating what a uniform Kan fibration is. If $p : Y \to X$ is a map, then we can define simplicial sets:

$$\begin{aligned} \mathrm{Probl}(p)_n &= \{(\pi : \Delta^n \times \mathbb{I} \to X, S \subseteq \Delta^n, y \in Y_n, \rho : S \times \mathbb{I} \to Y) : \\ &\qquad \pi(-, i) = p(y), p \circ \rho = \pi \restriction S \times \mathbb{I}, \rho(-, i) = y \restriction S\} \\ \mathrm{Sol}(p)_n &= \{(\pi, S, y, \rho, \lambda : \Delta^n \times \mathbb{I} \to Y) : \\ &\qquad (\pi, S, y, \rho) \in \mathrm{Probl}(p)_n, p \circ \lambda = \pi, \lambda(-, i) = y, \lambda \restriction S \times \mathbb{I} = \rho\} \end{aligned}$$

There is a forgetful map of simplicial maps $U_p : \mathrm{Sol}(p) \to \mathrm{Probl}(p)$. Saying that p has right lifting property against all maps of the form $m \hat{\otimes} \delta_i$ (with m a cofibrant sieve) is the same thing as saying that this map U_p is epi. For p to be a uniform

Kan fibration means that this map has a section: indeed, a uniform Kan fibration structure on p can be understood as a section of this map.

At this point it is good to remember that in a classical metatheory sections of a map $f : B \to A$ of simplicial sets can be constructed by induction on the dimension. That is, we construct maps $s_n : A_n \to B_n$ by induction on n, where we make a case distinction on whether an element $a \in A_n$ is degenerate or not. Indeed, if $a \in A_n$ is degenerate, then $a = a' \cdot \sigma$ for some unique non-degenerate a' and epi $\sigma : [n] \to [k]$. This means that the value $s_n(a)$ is determined by $s_k(a')$. If a is non-degenerate, then any face of $s(a)$ is already determined s_{n-1}; this means that $s(a)$ has to get the correct boundary as determined by s_{n-1} and be such that $p(s(a)) = a$. However, beyond these requirements we are free to choose $s(a)$ in any manner we like; meaning that any such choice will yield a map of simplicial sets. In particular, if we choose our solutions following this recipe the lifts will automatically satisfy the uniformity condition.

Note that a notion of fibred structure Fib could fail to be local in two different ways:

(1) There is a map $p : Y \to X$ and distinct elements $a, b \in \mathrm{Fib}(p)$ such that for any pullback square σ of the form

$$\begin{array}{ccc} Y_x & \longrightarrow & Y \\ \downarrow & & \downarrow p \\ \Delta^n & \xrightarrow{x} & X \end{array}$$

we have $\mathrm{Fib}(\sigma)(a) = \mathrm{Fib}(\sigma)(b)$.

(2) There is a map $p : Y \to X$ as well as a function a which assigns to each $x \in X_n$ an element $a_x \in \mathrm{Fib}(Y_x \to \Delta^n)$ in such a way that for any pullback square τ of the form

$$\begin{array}{ccc} Y_{x \cdot \alpha} & \longrightarrow & Y_x \\ \downarrow & & \downarrow \\ \Delta^m & \xrightarrow{\alpha} & \Delta^n \end{array}$$

we have $\mathrm{Fib}(\tau)(a_x) = a_{x \cdot \alpha}$, without there being an element $a \in \mathrm{Fib}(p)$ such that for any pullback square σ as in (1) we have $\mathrm{Fib}(\sigma)(a) = a_x$.

It turns out that the notion of a uniform Kan fibration fails to be local in both ways.

Proposition D.1 *The notion of a uniform Kan fibration fails to be local for the reason explained in (1).*

Proof Let $X = \Delta^1 \times \Delta^1$ and $Y = \mathrm{Cosk}^1(X \cup \{f\})$; that is, Y is the 1-coskeleton of $X \cup \{f\}$ where f is an additional copy of the edge $e : (1, 0) \to (1, 1)$. The map $p : Y \to X$ is the unique extension of the identity on X.

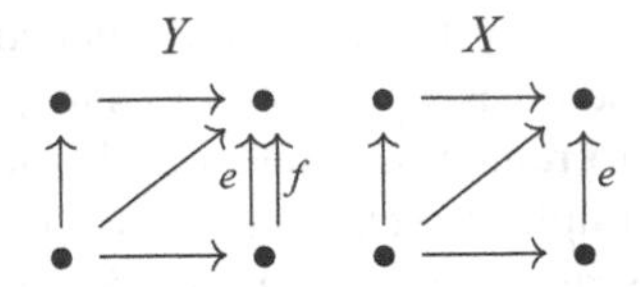

We wish to construct two uniform Kan fibration structures on p, which we do following the recipe outlined above. So suppose we have a lifting problem as follows:

$$\begin{array}{ccc} \Delta^n \times \{i\} \cup S \times \mathbb{I} & \longrightarrow & Y \\ \downarrow & & \downarrow p \\ \Delta^n \times \mathbb{I} & \longrightarrow & X. \end{array}$$

Note for any $x \in X_n$ we have:

$$p_n^{-1}(x) = \begin{cases} \{e, f\} & \text{if } x = e \\ \{\bullet\} & \text{else} \end{cases}$$

Therefore the number of solutions to a lifting problem as the one above is typically 1, except when the image of the map $\Delta^n \times \mathbb{I} \to X$ includes the edge e, while the image of the map $\Delta^n \times \{i\} \cup S \times \mathbb{I} \to X$ excludes it: in this case the number of solutions is 2, where each such is completely determined by the lift of the edge e (whether it is e or f).

So to define our uniform Kan fibration structures a, b we assume we are given a lifting problem as above where the image of the map $\Delta^n \times \mathbb{I} \to X$ includes the edge e, while the image of the map $\Delta^n \times \{i\} \cup S \times \mathbb{I} \to X$ excludes it. We do this by making a case distinction on whether the map $\Delta^n \times \mathbb{I} \to X$ factors through u where $u : \Delta^2 \to X$ if the unique inclusion whose image includes e. If it does, both a and b choose the lift which map e to e; if it does not, a lifts e to e, while b lifts e to f.

First of all, note that a and b are distinct, because they provide different solutions to problem of filling the "open box"

$$\begin{array}{ccc} \Delta^1 \times \{0\} \cup \{0, 1\} \times \mathbb{I} & \longrightarrow & Y \\ \downarrow & & \downarrow p \\ \Delta^1 \times \mathbb{I} & \longrightarrow & X \end{array}$$

where the map at the bottom is the identity.

It remains to check that for any $x : \Delta^n \to X$ both a and b induce the same uniform Kan fibration structure on $Y_x \to \Delta^n$. If e is not in the image of x, then this is clear. Otherwise there is a map $\sigma : \Delta^n \to \Delta^2$ such that $u\sigma = x$. Thus, it suffices to check that a and b pull back to the same element in $\mathrm{Fib}(Y_u \to \Delta^2)$. But this is clear by construction. □

Proposition D.2 *The notion of a uniform Kan fibration fails to be local for the reason explained in (2).*

Proof Let E and F be 3-simplices glued together along inclusions $\Delta^1 \times \Delta^1 \to S$ and $\Delta^1 \times \Delta^1 \to T$ to yield a simplicial set X. We name an edge $e : (1, 0) \to (1, 1)$. For Y we again take the 1-coskeleton of $X \cup \{f\}$ where f is an additional copy of the edge e and $p : Y \to X$ is the unique map extending the identity on X. Let Y_E and Y_F be the restriction of Y to E and F, respectively.

$$\begin{array}{ccc} Y_E & \longrightarrow & Y \\ \downarrow & & \downarrow p \\ E & \longrightarrow & X \end{array} \qquad \begin{array}{ccc} Y_F & \longrightarrow & Y \\ \downarrow & & \downarrow p \\ F & \longrightarrow & X \end{array}$$

We will define uniform Kan fibration structures a and b on $Y_E \to E$ and $Y_F \to F$, respectively, which restrict to the same uniform Kan fibration structure on any triangle in their overlap $\Delta^1 \times \Delta^1 \to X$. In doing so we will prove that any pullback of p along a representable carries a uniform Kan fibration structure, where these uniform Kan fibration structures are compatible as made precise in (2). We build these uniform Kan fibration structures as before, where for $Y_E \to E$ the fibration structure a will always lift e to e if possible. For $Y_F \to F$ the fibration structure b will always lift e to f if possible; however, if the lifting problem factors through a triangle of $\Delta^1 \times \Delta^1 \to X$ the fibration structure b will lift e to e. In this way we ensure that the uniform Kan fibration structures on $Y_E \to E$ and $Y_F \to F$ restrict to the same uniform Kan fibration structure on any triangle of their overlap $\Delta^1 \times \Delta^1 \to X$.

However, there is no uniform Kan fibration structure on p which restrict to both a and b. Indeed, there is open box filling problem on the overlap $\Delta^1 \times \Delta^1 \to X$ on which a and b want to lift e to e and f, respectively. □

References

1. K. Kapulkin, P.L. Lumsdaine, J. Eur. Math. Soc. **23**(6), 2071 (2021). https://doi.org/10.4171/JEMS/1050
2. M. Hofmann, T. Streicher, in *Twenty-Five Years of Constructive Type Theory (Venice, 1995).* Oxford Logic Guides, vol. 36 (Oxford University Press, New York, 1998), pp. 83–111
3. S. Awodey, M.A. Warren, Math. Proc. Camb. Philos. Soc. **146**(1), 45 (2009). https://doi.org/10.1017/S0305004108001783
4. B. van den Berg, R. Garner, Proc. Lond. Math. Soc. (3) **102**(2), 370 (2011). https://doi.org/10.1112/plms/pdq026
5. P.L. Lumsdaine, Log. Methods Comput. Sci. **6**(3), 3:24, 19 (2010). https://doi.org/10.2168/LMCS-6(3:24)2010
6. T.U.F. Program, *Homotopy Type Theory—Univalent Foundations of Mathematics* (The Univalent Foundations Program, Princeton, NJ; Institute for Advanced Study (IAS), Princeton, NJ, 2013)
7. M. Bezem, T. Coquand, E. Parmann, in *13th International Conference on Typed Lambda Calculi and Applications, LIPIcs. Leibniz Int. Proc. Inform.*, vol. 38 (Schloss Dagstuhl. Leibniz-Zent. Inform., Wadern, 2015), pp. 92–106
8. N. Gambino, C. Sattler, J. Pure Appl. Algebra **221**(12), 3027 (2017). https://doi.org/10.1016/j.jpaa.2017.02.013
9. C. Cohen, T. Coquand, S. Huber, A. Mörtberg, FLAP **4**(10), 3127 (2017). http://collegepublications.co.uk/ifcolog/?00019
10. M. Hofmann, T. Streicher, Lifting Grothendieck universes (1997). Available from https://www2.mathematik.tu-darmstadt.de/~streicher/NOTES/lift.pdf
11. P. Aczel, M. Rathjen, Notes on constructive set theory. Tech. Rep. No. 40, Institut Mittag-Leffler (2000/2001)
12. T. Coquand, S. Huber, C. Sattler, in *4th International Conference on Formal Structures for Computation and Deduction, FSCD 2019, June 24-30, 2019, Dortmund, Germany*. LIPIcs, vol. 131, ed. by H. Geuvers (Schloss Dagstuhl - Leibniz-Zentrum für Informatik, 2019), pp. 11:1–11:23. DOI 10.4230/LIPIcs.FSCD.2019.11. https://doi.org/10.4230/LIPIcs.FSCD.2019.11
13. S. Huber, J. Autom. Reason. **63**(2), 173 (2019). https://doi.org/10.1007/s10817-018-9469-1
14. M. Bezem, T. Coquand, S. Huber, J. Autom. Reason. **63**(2), 159 (2019). https://doi.org/10.1007/s10817-018-9472-6
15. S. Henry, A constructive account of the Kan-Quillen model structure and of Kan's Ex^∞ functor (2019). arXiv:1905.06160

B. van den Berg, E. Faber, *Effective Kan Fibrations in Simplicial Sets*,
Lecture Notes in Mathematics 2321, https://doi.org/10.1007/978-3-031-18900-5

16. N. Gambino, S. Henry, J. Lond. Math. Soc. (2) **105**(2), 1073 (2022). https://doi.org/10.1112/jlms.12532
17. N. Gambino, C. Sattler, K. Szumiło, Quart. J. Math. (2022). https://doi.org/10.1093/qmath/haab057
18. P. Gabriel, M. Zisman, *Calculus of Fractions and Homotopy Theory*. Ergebnisse der Mathematik und ihrer Grenzgebiete, Band 35 (Springer Inc., New York, 1967)
19. B. van den Berg, R. Garner, ACM Trans. Comput. Log. **13**(1), 44 (2012). Art. 3. https://doi.org/10.1145/2071368.2071371
20. I. Orton, A.M. Pitts, Log. Methods Comp. Sci. **14**(4) (2018). https://doi.org/10.23638/LMCS-14(4:23)2018
21. D. Frumin, B. van den Berg, Math. Struct. Comput. Sci. **29**(4), 588 (2019). https://doi.org/10.1017/S0960129518000142
22. E. Riehl, New York J. Math. **17**, 173 (2011). http://nyjm.albany.edu:8000/j/2011/17_173.html
23. J. Bourke, R. Garner, J. Pure Appl. Algebra **220**(1), 108 (2016). https://doi.org/10.1016/j.jpaa.2015.06.002
24. J. Bénabou, J. Roubaud, C. R. Acad. Sci. Paris Sér. A-B **270**, A96 (1970)
25. P.T. Johnstone, *Sketches of an Elephant: A Topos Theory Compendium, vol. 1*. Oxford Logic Guides, vol. 43 (The Clarendon Press Oxford University Press, New York, 2002)
26. C. Sattler, The equivalence extension property and model structures (2018). arXiv:1704.06911
27. D.C. Cisinski, Univalent universes for elegant models of homotopy types (2014). arXiv:1406.0058
28. M. Shulman, All $(\infty, 1)$-toposes have strict univalent universes (2019). arXiv:1904.07004
29. J. Beck, in *Sem. on Triples and Categorical Homology Theory (ETH, Zürich, 1966/67)* (Springer, Berlin, 1969), pp. 119–140
30. J. Power, H. Watanabe, in *Coalgebraic Methods in Computer Science (Amsterdam, 1999)*, vol. 280 (Elsevier, 2002), pp. 137–162. https://doi.org/10.1016/S0304-3975(01)00024-X
31. R. Garner, Appl. Categ. Struct. **17**(3), 247 (2009). https://doi.org/10.1007/s10485-008-9137-4
32. G. Rosolini, Continuity and effectiveness in topoi. Ph.D. thesis, Carnegie Mellon University (1986). ftp://ftp.disi.unige.it/pub/person/RosoliniG/papers/coneit.ps.gz
33. P. North, Type-theoretic weak factorization systems (2019). arXiv:1906.00259
34. D.C. Cisinski, *Higher Categories and Homotopical Algebra*, vol. 180. Cambridge Studies in Advanced Mathematics (Cambridge University Press, Cambridge, 2019). https://doi.org/10.1017/9781108588737
35. J. Lurie, *Higher Topos Theory*, vol. 170. Annals of Mathematics Studies (Princeton University Press, Princeton, NJ, 2009). https://doi.org/10.1515/9781400830558
36. J.P. May, *Simplicial Objects in Algebraic Topology*, vol. 11. Van Nostrand Mathematical Studies (D. Van Nostrand Co., Inc., Princeton, NJ, Toronto, ON, London, 1967)
37. C. Berger, Une version effective du théorème de Hurewicz. Ph.D. thesis, Grenoble (1991)
38. C. Berger, Bull. Soc. Math. France **123**(1), 1 (1995). http://www.numdam.org/item?id=BSMF_1995__123_1_1_0
39. T. Von Glehn, Polynomials and models of type theory. Ph.D. thesis, University of Cambridge (2015). https://doi.org/10.17863/CAM.16245
40. M.G. Abbott, T. Altenkirch, N. Ghani, in *Foundations of Software Science and Computational Structures, 6th International Conference, FOSSACS 2003 Held as Part of the Joint European Conference on Theory and Practice of Software, ETAPS 2003, Warsaw, Poland, April 7–11, 2003, Proceedings* (2003), pp. 23–38. https://doi.org/10.1007/3-540-36576-1_2
41. N. Gambino, J. Kock, Math. Proc. Camb. Philos. Soc. **154**(1), 153 (2013). https://doi.org/10.1017/S0305004112000394
42. T. Streicher, Fibred categories à la Jean Benabou (2018). arXiv:1801.02927
43. D. Ahman, T. Uustalu, in *Proceedings of the 6th International Workshop on Bidirectional Transformations co-located with The European Joint Conferences on Theory and Practice of Software, BX@ETAPS 2017, Uppsala, Sweden, April 29, 2017, CEUR Workshop Proceedings*, vol. 1827, ed. by R. Eramo, M. Johnson (CEUR-WS.org, 2017), *CEUR Workshop Proceedings*, vol. 1827, pp. 59–73. http://ceur-ws.org/Vol-1827/paper11.pdf

44. E. Faber, Homogenous models and their toposes of supported sets. Ph.D. thesis, University of Cambridge (2019). https://doi.org/10.17863/CAM.49282
45. I. Moerdijk, E. Palmgren, in *Proceedings of the Workshop on Proof Theory and Complexity, PTAC'98 (Aarhus)*, vol. 104(1–3) (2000), pp. 189–218. https://doi.org/10.1016/S0168-0072(00)00012-9
46. B. van den Berg, I. Moerdijk, Math. Struct. Comput. Sci. **25**(5), 1100 (2015). https://doi.org/10.1017/S0960129514000516
47. C. Sattler, Failure of the algebraic universe of types with prism-based Kan composition in simplicial sets. Unpublished note dated 15 August 2018

Index

B. van den Berg, E. Faber, *Effective Kan Fibrations in Simplicial Sets*,
Lecture Notes in Mathematics 2321, https://doi.org/10.1007/978-3-031-18900-5

LECTURE NOTES IN MATHEMATICS Springer

Editors in Chief: J.-M. Morel, B. Teissier;

Editorial Policy

1. Lecture Notes aim to report new developments in all areas of mathematics and their applications – quickly, informally and at a high level. Mathematical texts analysing new developments in modelling and numerical simulation are welcome.

   Manuscripts should be reasonably self-contained and rounded off. Thus they may, and often will, present not only results of the author but also related work by other people. They may be based on specialised lecture courses. Furthermore, the manuscripts should provide sufficient motivation, examples and applications. This clearly distinguishes Lecture Notes from journal articles or technical reports which normally are very concise. Articles intended for a journal but too long to be accepted by most journals, usually do not have this "lecture notes" character. For similar reasons it is unusual for doctoral theses to be accepted for the Lecture Notes series, though habilitation theses may be appropriate.

2. Besides monographs, multi-author manuscripts resulting from SUMMER SCHOOLS or similar INTENSIVE COURSES are welcome, provided their objective was held to present an active mathematical topic to an audience at the beginning or intermediate graduate level (a list of participants should be provided).

   The resulting manuscript should not be just a collection of course notes, but should require advance planning and coordination among the main lecturers. The subject matter should dictate the structure of the book. This structure should be motivated and explained in a scientific introduction, and the notation, references, index and formulation of results should be, if possible, unified by the editors. Each contribution should have an abstract and an introduction referring to the other contributions. In other words, more preparatory work must go into a multi-authored volume than simply assembling a disparate collection of papers, communicated at the event.

3. Manuscripts should be submitted either online at www.editorialmanager.com/lnm to Springer's mathematics editorial in Heidelberg, or electronically to one of the series editors. Authors should be aware that incomplete or insufficiently close-to-final manuscripts almost always result in longer refereeing times and nevertheless unclear referees' recommendations, making further refereeing of a final draft necessary. The strict minimum amount of material that will be considered should include a detailed outline describing the planned contents of each chapter, a bibliography and several sample chapters. Parallel submission of a manuscript to another publisher while under consideration for LNM is not acceptable and can lead to rejection.

4. In general, **monographs** will be sent out to at least 2 external referees for evaluation.

   A final decision to publish can be made only on the basis of the complete manuscript, however a refereeing process leading to a preliminary decision can be based on a pre-final or incomplete manuscript.

   Volume Editors of **multi-author works** are expected to arrange for the refereeing, to the usual scientific standards, of the individual contributions. If the resulting reports can be forwarded to the LNM Editorial Board, this is very helpful. If no reports are forwarded or if other questions remain unclear in respect of homogeneity etc, the series editors may wish to consult external referees for an overall evaluation of the volume.

5. Manuscripts should in general be submitted in English. Final manuscripts should contain at least 100 pages of mathematical text and should always include

   – a table of contents;
   – an informative introduction, with adequate motivation and perhaps some historical remarks: it should be accessible to a reader not intimately familiar with the topic treated;
   – a subject index: as a rule this is genuinely helpful for the reader.
   – For evaluation purposes, manuscripts should be submitted as pdf files.

6. Careful preparation of the manuscripts will help keep production time short besides ensuring satisfactory appearance of the finished book in print and online. After acceptance of the manuscript authors will be asked to prepare the final LaTeX source files (see LaTeX templates online: https://www.springer.com/gb/authors-editors/book-authors-editors/manuscriptpreparation/5636) plus the corresponding pdf- or zipped ps-file. The LaTeX source files are essential for producing the full-text online version of the book, see http://link.springer.com/bookseries/304 for the existing online volumes of LNM). The technical production of a Lecture Notes volume takes approximately 12 weeks. Additional instructions, if necessary, are available on request from lnm@springer.com.

7. Authors receive a total of 30 free copies of their volume and free access to their book on SpringerLink, but no royalties. They are entitled to a discount of 33.3 % on the price of Springer books purchased for their personal use, if ordering directly from Springer.

8. Commitment to publish is made by a *Publishing Agreement*; contributing authors of multiauthor books are requested to sign a *Consent to Publish form*. Springer-Verlag registers the copyright for each volume. Authors are free to reuse material contained in their LNM volumes in later publications: a brief written (or e-mail) request for formal permission is sufficient.

Addresses:

Professor Jean-Michel Morel, CMLA, École Normale Supérieure de Cachan, France
E-mail: moreljeanmichel@gmail.com

Professor Bernard Teissier, Equipe Géométrie et Dynamique,
Institut de Mathématiques de Jussieu – Paris Rive Gauche, Paris, France
E-mail: bernard.teissier@imj-prg.fr

Springer: Ute McCrory, Mathematics, Heidelberg, Germany,
E-mail: lnm@springer.com

Printed in the United States
by Baker & Taylor Publisher Services

Printed in the United States
by Baker & Taylor Publisher Services